ACCESS TO JUSTICE, DIGITALIZATION AND VULNERABILITY

Perspectives on
Law and Access to Justice

Series Editors: **Jess Mant**, Monash University and
Daniel Newman, Cardiff University

This series forges a coherent field of access to justice scholarship, drawing connections between research from different disciplines such as civil and criminal law, family law, housing law, immigration law and social welfare law. It facilitates a multifaceted critique of the key issues and incorporates a diverse range of perspectives that will shape the future directions of this emerging field.

International advisory board:

Find out more at

bristoluniversitypress.co.uk/
perspectives-on-law-and-access-to-justice

ACCESS TO JUSTICE, DIGITALIZATION AND VULNERABILITY

Exploring Trust in Justice

Naomi Creutzfeldt, Arabella Kyprianides,
Ben Bradford and Jonathan Jackson

With a Foreword by
Jess Mant and Daniel Newman

BRISTOL
UNIVERSITY
PRESS

First published in Great Britain in 2024 by

Bristol University Press
University of Bristol
1–9 Old Park Hill
Bristol
BS2 8BB
UK
t: +44 (0)117 374 6645
e: bup-info@bristol.ac.uk

Details of international sales and distribution partners are available at bristoluniversitypress.co.uk

© Bristol University Press 2024

British Library Cataloguing in Publication Data
A catalogue record for this book is available from the British Library

ISBN 978-1-5292-2952-3 hardcover
ISBN 978-1-5292-2953-0 ePub
ISBN 978-1-5292-2954-7 ePdf

The right of Naomi Creutzfeldt, Arabella Kyprianides, Ben Bradford and Jonathan Jackson to be identified as authors of this work has been asserted by them in accordance with the Copyright, Designs and Patents Act 1988.

Cover design: Nicky Borowiec
Front cover image: Adobe Stock/pandaclub23
Bristol University Press uses environmentally responsible print partners.
Printed and bound in Great Britain by CPI Group (UK) Ltd, Croydon, CR0 4YY

FSC
www.fsc.org
MIX
Paper | Supporting responsible forestry
FSC® C013604

Contents

Series Editors' Preface

Jess Mant (Monash University) and
Daniel Newman (Cardiff University)

In recent years, there has been increasing global interest in issues of access to justice and how these issues relate to our increasingly unequal societies. The concept of 'access to justice' – the protection of the law that enables citizens to enforce their rights – is far from new, but has often been approached from a range of different perspectives, methodologies and disciplines. This series provides a home for socio-legal scholarship which explores issues of access to justice. It will draw important connections between research undertaken across different sub-disciplines which intrinsically involve issues of access to justice but are not routinely in conversation with one another, such as civil and criminal law, family law, housing law, immigration law and social welfare law. The series will forge a coherent field of access to justice scholarship; incorporating a diverse range of perspectives to facilitate a multifaceted critique of the key issues, debates and challenges that underpin access to justice, and shape the future directions of this emerging field.

List of Figures and Tables

Figures

Tables

About the Authors

Naomi Creutzfeldt is Professor of Law and Society at the University of Kent. She teaches public law, socio-legal theory and methods, and AI in dispute resolution (at the University of Oxford). Her interests in administrative and civil justice systems and alternative dispute resolution (as pathways of informal dispute resolution) have a broader scope, addressing questions of access to justice, vulnerability, digitalization and consumer protection.

Arabella Kyprianides is Research Fellow at the UCL Institute of Security and Crime Science and an active member of the UCL Centre for Global City Policing and the Keele Policing Academic Collaboration. Arabella is involved in ESRC and Nuffield-funded research grants, and various consultancy projects in areas of intervention evaluation, policing, crime and recidivism. Her research interests include public trust, police legitimacy, and compliance in the context of policing marginalized communities; as well as social determinants of wellbeing among vulnerable groups, such as social identity.

Ben Bradford is Professor of Global City Policing at the Department of Security and Crime Science, where he is also Director of the JDI Centre for Global City Policing. His research interests include procedural justice theory, trust and legitimacy in policing and justice settings, and questions of cooperation and compliance. He has also published on various aspects of 'public-facing' police work, the use of force, and ethnic/racial disproportionality.

Jonathan Jackson is Professor of Research Methodology at the LSE. He is an Honorary Professor of Criminology at the University of Sydney Law School and an Affiliated Scholar in the Justice Collaboratory of Yale Law School.

Acknowledgements

We are grateful to the Nuffield Foundation for funding our project 'Delivering Administrative Justice after the Pandemic',[1] on which this book is based. The focus of our book is on ombuds, tribunals and advice providers as pathways to justice for people with problems in the areas of housing and special educational needs and disabilities (SEND). We are most grateful to those who have taken the time to be interviewed by us and to share with us their experiences of (non) access to the digital justice system. We are fortunate to have had support from professionals administering the pathways we investigate, who shared with us their experiences of providing their services during and after the pandemic. This book is truly enriched by the privilege of presenting and discussing a diverse range of voices.

We are grateful to Helen Davies and Grace Carroll from BUP for their support and guidance, as well as to the wonderful series editors Jess Mant and Daniel Newman for accepting us as their first book. A special thanks to Heidi Bancroft who was a vital part of our research project team. We also thank Tara Mulqueen, Matt Howard, Clare Williams and Ian Loader for comments on selected chapters and ideas. We are most grateful to the anonymous reviewer for helpful suggestions on how to improve the manuscript. As always, huge thanks to Marie Selwood for editing this book.

Note
[1]   www.nuffieldfoundation.org/project/delivering-administrative-justice-after-the-pandemic

Foreword

Jess Mant (Monash University, Australia) and
Daniel Newman (Cardiff University, UK)

We are delighted to have been invited to write the foreword for this book; not only because it has been authored by an eminent team of academics working across our specialisms, but because it speaks directly to topics that we believe are vitally important for access to justice scholarship. This book is about understanding the extent to which people were able to access the administrative justice system of England and Wales during the COVID-19 pandemic, and what their experiences can tell us about the future of access to justice as society moves forward beyond the initial impact of the pandemic. As researchers who both work in the access to justice field, for some time we have been concerned about the extent to which members of the public can access, use and meaningfully participate in justice systems, as well as how they can best be supported to recognize, address and resolve problems for which there are potential legal remedies or consequences. Since the outbreak of COVID-19, we have undertaken various projects together which have been geared towards understanding how this global event – and all of its many ramifications – have impacted people and their relationship with law to date, as well as what the legacy of the pandemic may be for those pursuing access to justice in the future. It is no surprise, therefore, that we were delighted to have the opportunity to engage with the rich and varied evidence on offer in this book, and that we view it as a valuable resource for those working in and around issues of justice.

The book is timely and important in its focus on the notion of digitalized justice and the administrative justice system. In recent years, the COVID-19 pandemic has marked a seismic shift across all areas of this system; it has forced judges, litigants, lawyers, advice-seekers and advice providers alike to adapt to digital methods of communication, virtual hearings and remote advice-giving. These rapidly accelerated trends towards digitalization were already moving – albeit slowly – in some areas of law, while forcing radical new ways of working on others. While there has been a succession of important reports on the increased digitalization of justice following the pandemic, this is the first book-length treatment of the subject, and the first to draw explicit connections between pre and post-pandemic implications of digital justice. In response to this evidence gap, the authors provide a much-needed empirical and theoretical contribution to our understanding of how legal institutions operate in the areas of housing and special educational needs and disability (SEND). Most importantly, this contribution is rooted in a comprehensive dataset comprising real-life experiences of legal (and non-legal) systems as well as first-hand perceptions of justice.

The book is unique in that it presents the findings of a large-scale, mixed-methods study. In our experience, there are typically two core communities of thought within the fields of socio-legal studies and access to justice. On one hand, there are those who advocate for large-scale, representative data that can reveal comprehensive insights into how people use (or do not use) legal institutions to resolve their legal problems. On the other hand, there are scholars who espouse the importance of learning about the detailed stories and journeys of users of these systems through 'bottom-up' approaches. Evidence in both regards is important, but in many empirical socio-legal projects, there can certainly be pressure to devote research efforts fully to one of these two very different concerns. In light of this, it is both refreshing and rare to see a book that seeks to ask questions – and more importantly, provide answers – that speak to both of these scholarly positions. The authors do so by combining data drawn from an extensive number of qualitative interviews, which can tell us much about the range of experiences and challenges occurring within and beyond legal institutions, with an innovative quantitative experiment, from which we can learn about the widespread extent of trust that members of the general public place in legal systems. By combining these research methods, the authors are able to provide a wide-ranging assessment of how effectively legal institutions understand and respond to the needs of their users ('help-seekers'), which examines the legal system from all angles. As the authors explain, our justice institutions are largely 'disconnected' and 'operate in silos', which further complicates the task of unravelling the impacts of digitalization and the COVID-19 pandemic. As such, the exhaustiveness of this approach has never been more necessary.

This book also comes at a crucial time for legal professionals and those who work in and for the justice system, as they begin to come to terms with what comes next for digital (and non-digital) justice. It is also a pivotal moment for governments and legal institutions, as they begin to digest the lessons that have been learnt during the pandemic. There are now important decisions to be made about what has worked well – which innovations can be taken forward, improved or built upon – and what was problematic – which pandemic-necessitated practices need to be abandoned or rethought now that in-person options are possible again. While there are myriad questions of this nature, this book provides a clear response: that access to justice requires legal institutions to cater to all members of society, 'not just those who are legally and technically capable'. In making this argument, the authors elucidate that access to justice is far from a pandemic-specific problem. Barriers to justice do not only come in the form of internet access or digital literacy. Rather, our legal institutions have been characterized by historic and longstanding concerns about the needs of lay users – from the traditional architectural design of courthouses, to the unwavering assumption that people are able to easily identify the legal components of their problems, or compartmentalize their emotional responses to personal crises. As the authors argue, while it is important to consider the impact of the pandemic on access to justice, it is imperative to recognize that the extent to which people place trust in justice more broadly comprises many other factors that preceded the first lockdown in March 2020.

In our view, there are two key contributions of this book which will significantly influence current understandings of access to justice. The first is the relationship between access to justice and vulnerability, and how this informs our understanding of legal capability. To some extent within policy literature, the notion of vulnerability has been used pejoratively as a means of identifying specific populations as – at best – part of a limited group designated to be deserving of legal assistance which is being curtailed for budgetary reasons, or – at worst – deficient in legal capability, meaning that their autonomy and participation within legal processes should be minimized. Even within legal scholarship, the notion of vulnerability can frequently be used in a categorical sense to identify individuals who may struggle to access or use law. As the authors acknowledge: 'Individuals who are vulnerable, such as those who are on a low income, have disabilities, or are part of marginalized communities, may face obstacles in accessing advice, legal services and navigating the justice system.'

However, these simplistic equations of vulnerability with certain circumstances or personal characteristics can, as the authors argue, mask the significant impact of *hidden* vulnerabilities:

> [H]idden vulnerabilities (factors that are not immediately apparent) can pose barriers to accessing justice. These factors can include poverty, lack of education, mental health issues and social marginalization. Those who are affected by hidden vulnerabilities might have problems in navigating the justice system (for example, understanding the process, fear of retaliations, being prone to give up or not even try). (p 200)

As researchers, we have personally grappled with the concept of vulnerability over many years, and have argued many times – in alignment with others who have been equally inspired by the work of Martha Fineman – that vulnerability is not simply a personal characteristic but a universal human experience. In other words, there is no such thing as 'invulnerability'; rather, the differences between us relate more clearly to the specific support needs that we all have at any given time during our lifetimes. As we have argued elsewhere, when it is accepted that all citizens, at one stage or another, will require some degree of support, and that the state has a central role in facilitating that support, we can begin the more productive task of identifying ways to target bespoke assistance to those who need it, when they need it. We were delighted to see the authors engage with vulnerability in a similar vein in this book. As they argue, an individual's experience of vulnerability within the legal system varies significantly according to several contextual factors, including but not limited to: their level of technological competence; the extent to which they are able to access and use non-legal community resources; the trust they place in the legal system; and their prior experiences with discrimination, marginalization or bias in respect of public institutions. All of this, the authors argue, can be used to expand our understandings of 'digital legal consciousness' – and to this end, they set out precise guidelines for how access to justice scholars can look beyond both digital and legal capabilities frameworks in their evaluations of new justice innovations and initiatives.

The second contribution is the way that this book centres the personal nature of the journey that people take when they experience a justice problem. There is a great deal of access to justice literature that focuses on the journeys that people take through legal systems, and for good reason: there is much to learn at each stage of a justice journey. First, there are factors that weigh into people's decisions and behaviours when they first experience a legal problem. Next, there are variances in how they may respond when they face different challenges or barriers to seeking help and support. Finally, there are differences in how they engage with formal processes of the legal system. This book builds upon this by exploring the significant impact of *emotion* on these trajectories. In identifying serious concerns about public trust in the legal system, the authors consider the potential benefit of initiatives that are geared towards promoting positive perceptions of legal services, such as outreach campaigns or community education initiatives. This serves as yet another reminder that barriers to justice are not merely practical, and are certainly not limited to technological barriers. Rather, accessibility barriers can be invisible, cultural and social. Appreciating this, we can garner insight into the scale of the work that is to be done if we are to strive towards a legal system that is truly accessible and open to all. Nevertheless, the work required is inevitably even greater than we imagine because, as the authors acknowledge, sadly: 'People's negative perceptions of services are often well founded. Therefore, one aim has to be to encourage people to see that there may be some worthwhile possibility of justice so that they will consider taking action.'

In other words, it is not just a case of giving people the practical tools they need to participate in justice, and it is also not merely a case of encouraging people to recognize legal systems as sources of support that they can trust and rely on. Rather, what is needed is a wholescale transformation of the lived realities that people are having within these systems. From this perspective, an access to justice journey is far from linear. Mistrust of law is bred through repeated negative experiences, and as a community of access to justice scholars, we must do more to acknowledge the important role that emotion can play at every stage of the different journeys that people take in response to, and in pursuit of, justice.

To this end, the authors deliver a strong argument for an expanded definition of access to justice: one that extends beyond legal definitions of justice (assistance, advice and resolutions obtained through legal institutions), and instead recognizes the potential significance of community-led initiatives and other non-legal forms of support. After all, it is only by broadening our view beyond law that we can properly assess the efficacy of legal processes, methodologies and solutions, and begin to think more holistically about ways to transform lived experiences of justice.

Finally, we would like to note (with great excitement) that this is the first book in our co-edited series with Bristol University Press, *Perspectives on Law and Access to Justice*. We could not be more delighted to be launching our series with this book. It is exactly the kind of compelling and innovative access to justice scholarship that we envisaged in the early stages of cultivating this series as a pioneering space for interdisciplinary, empirical and policy-focused access to justice research. This book directly engages with several of the key themes that underpin the series: it develops theories of access to justice by using legal consciousness to underpin the understanding of the sector; it explores interaction with the institutions of justice in exploring how trust and legitimacy of ombuds, tribunals and advice provision are impacted by digitalization; and it considers legal need by exploring the factors that may facilitate disengagement with the justice system. A key motivation for us – as scholars who felt 'siloed' at early stages of our careers within family law and criminal law, respectively – was to create a series where authors would not only be free to take interdisciplinary approaches to their work, but also feel empowered to overcome traditional *intra*-disciplinary boundaries of legal scholarship. In bringing together the distinct areas of housing and SEND, and contextualizing these problems within the broader legal system, this book provides a pioneering example of how to achieve this, as well as the significant benefits that this can have for the reach and impact of access to justice research. We hope that this inspires a wider community of scholars to feel less constrained by inter (and intra) disciplinary boundaries, and to engage in creative, innovative approaches to providing answers to the many unanswered questions and concerns about access to justice.

Introduction

This is a book about institutions in the administrative justice system (AJS), their users and pathways to justice. The AJS is made up of institutions that help individuals when the government acts in ways that are unfair or unjust (Adler 2003; Mullen 2010, 2016; Kirkham 2016; Tomlinson 2017b). The institutions that form the AJS are complaint schemes, ombuds,[1] tribunals and the Administrative Court. They influence our lives in areas of housing, health care, education, social security and taxation, for example.

The Administrative Justice and Tribunals Council established a framework for understanding the intricate connections between the decision-making process that underpins the relationship between the state and its citizens, as well as the methods utilized to resolve disputes, such as internal or external complaints and reviews, and the involvement of independent complaint handlers such as ombuds and tribunals (House of Commons 2023). As our book will show in later chapters, these associations are not actual connections between the institutions. Rather, the institutions are disconnected and typically operate in silos, which contributes to an over-complication of the AJS for those who work in these institutions as well as for people who seek access to these institutions.

This book is based on a Nuffield-funded research project[2] and is about people who administer institutions of the AJS, about people who use institutions of the AJS, and about those who do not access institutions of the AJS. We explore these different positions in the AJS through two distinct pathways to seek redress: housing and special educational needs and disabilities (SEND). We focus on these areas because they have been especially affected by two major changes to the justice system: digitalization (Tomlinson 2017a; Ryder 2019) and the COVID-19 pandemic (Creutzfeldt and Sechi 2021). The pathways through the justice system that we look at include advice services, non-governmental organizations (NGOs), ombuds and tribunals. The emphasis on these institutions allows us to better understand the effects of the pandemic; how such institutions managed to provide their services remotely; and how people accessed these services.

This book extends existing research by examining the effect of rapid digitalization on the delivery of justice. Lessons learned from delivering remote justice during the pandemic need to be evaluated and translated into practice. This includes documenting what works well and what can be changed to improve access for those further sidelined because of the pandemic. COVID-19 has forced the justice system, where possible, to go digital at a rapid pace. By empirically understanding areas that work well and those that need improvement, there is a huge opportunity to draw positive lessons from this crisis.

There have been some excellent projects that have sought to understand the impact of the pandemic on individual justice settings in recent years: for example, the family court (Ryan et al 2020), judicial review (Tomlinson et al 2020), video hearings (JUSTICE 2020), digital exclusion (Good Things Foundation 2020), the advice sector (Sechi 2020; Creutzfeldt and Sechi 2021) and the civil justice system (Legal Education Foundation 2019). Building on this, we explore the effects of rapid digitalization and the pandemic on the advice and redress system as well as its users; on access for marginalized groups; and on how trust can be built and sustained in specific parts of the AJS affected by the pandemic and digitalization. The following questions guided our research:

1. How is a siloed landscape of tribunals, ombuds, advice and NGOs able to provide access to justice, enacting values of respect, equality and accountability?
2. What lessons about digitalization and pathways to justice can be learned?
3. How can trust in justice more broadly – the belief that the justice system is fair, effective and open to all – be maintained?

Administrative justice is about ensuring correct decisions and fair process, but it is also about enacting values of respect, equality and deservingness – so that public bodies can be held to account. This book provides a detailed understanding of existing pathways to access the AJS in the two areas of law, and the reality of access. It explores the barriers that marginalized groups face to access justice and people's trust in online justice post-pandemic. We do this by bringing together, and contributing to, theories from different academic disciplines. We seek to advance scholarship around administrative justice (law), procedural justice (social psychology), access to justice (law and sociology), and legal consciousness (law, sociology, anthropology) through a rich theoretical (Part II) and empirical inquiry (Part III). Our theoretically informed mixed methodologies provided us with qualitative and quantitative datasets to draw upon to make sense of people's encounters with the AJS, as well as understand better why some did not reach it.

There is a lot to say about the developments in the AJS, especially in light of the court reform agenda and the COVID-19 pandemic. This book offers a start to discover how professionals have managed to deliver their services and how people have managed to access these services. To set the scene, in the following we will outline the two pathways we are focusing on in this book, the bodies that make up these pathways, and the context of the court reform. Then, we discuss our methodology and the outline of the book to show how all the parts fit together.

Two dispute resolution pathways and potential partnerships

We choose to look at two distinct dispute resolution pathways through the justice system; those for people with housing problems (Chapter 4) and those for people with SEND problems (Chapter 5). Within these pathways we are interested in people's journeys to seek advice, go to an ombuds (Housing Ombudsman, Local Government and Social Care Ombudsman (LGSCO), Parliamentary and Health Service Ombudsman (PHSO)), a tribunal (Property Chamber and SEND Tribunal), or to do nothing. The areas of housing and SEND have seen a rise in cases, especially during the pandemic. This, coupled with the impact of remote delivery of advice provision during the pandemic, has made access to justice even more of a challenge. It is crucial, therefore, to explore how the institutions providing advice and redress can work better together to increase access to justice for their users. Our data provide a unique opportunity to identify best practice as well as problems, in order to help build a better, more joined-up, system.

Initially, the project was driven by the prospect of supporting ombuds and tribunals – institutions that typically do not interact with each other – in creating a partnership (pre-pandemic) to enable signposting between each other (Kirkham and Creutzfeldt 2019; Creutzfeldt et al 2023): we revisit the idea in the conclusion. During the pandemic, the Housing Ombudsman and the Property Chamber started an informal referral system (after an Administrative Justice Council event introducing the idea in 2020) and both the SEND Tribunal and LGSCO are very keen to set up a Memorandum of Understanding and start a partnership (Creutzfeldt 2022). This would be a step towards joining up the AJS and making it easier to navigate for the help-seeker. The PHSO is interested in exploring a partnership model too. Therefore, we have included the PHSO in parts of our project. We hope that this book will provide some of the necessary information and evidence to support such partnership arrangements in the future.

Next, we briefly outline what ombuds and tribunals are, as well as the importance of the advice sector, before discussing the court modernization agenda.

The public ombuds

Ombuds are institutions that provide alternative dispute resolution (ADR). The ombuds model was introduced in Sweden (1809) as an institution to resolve citizens' complaints against the state. The ombuds model exists in most countries around the world (Seneviratne 2002; Buck et al 2016; Creutzfeldt 2022; Groves and Stuhmcke 2022). Ombuds also exist in the private sector, where they resolve disputes between consumers and businesses (Creutzfeldt 2018). Generally, ombuds have been set up to restore public confidence in administration (Creutzfeldt 2018). The ombuds model, due to its tremendous potential to process a high proportion of unmet legal needs for certain types of problems, draws its strength from its variety of contextual and conceptual adaptations (Carl 2012). Ombuds are free of charge for the citizen to use.

Tribunals

We discuss the SEND Tribunal and Property Chamber and their procedures in Part II. During the 2000s, tribunals turned from being quasi-judicial alternatives to become a specialist court through the Tribunals, Courts and Enforcement Act 2007 (Stebbings 2006; Drewry 2009; Carnwath 2011). Since then, there has been a strong growth in tribunals which has resulted in lengthy procedures that are time-consuming and costly, mainly due to austerity measures not being able to match the growth of the institutions (Drewry 2009; Adler 2012). Tribunals operate independently of the government, and there are debates about tribunals being more accessible for those who cannot afford legal representation (Genn 1993; Thomas 2016; Leader 2017; McKeever et al 2018).

Advice services

Throughout the project, as will feature in the help-seeker journeys in Part II, our data have demonstrated the importance of advice (Kirwan 2016; Creutzfeldt and Sechi 2021; Koch and James 2022). The advice sector provides free advice and support for people with problems in general or specialist areas. In our study we encountered those services which support people with housing or SEND matters. These advice providers have different funding structures, different specialties and must respond and adapt to their local community's needs (Mayo et al 2015). Especially during the pandemic this brought to the surface the real difficulties with technology, challenges of English not being a first language, and many other obstacles. Our interviews told a story of the advice sector being under-funded, under-staffed and constantly firefighting.

In sum, when we speak about a pathway for the help-seeker with a SEND or housing problem, we think of this process as starting with seeking advice (from trusted people and advice providers, for example), then accessing the formal processes of an ombuds or tribunal. These pathways are often complicated to enter and to navigate. The convergence of the court modernization programme and the unprecedented challenges posed by the COVID-19 pandemic has inadvertently created significant barriers within the pathways available to certain groups of help-seekers.

The court modernization (digitalization) programme

The pandemic has radically altered the landscape for processing any kind of administrative needs and disputing processes. Independent of the pandemic, there is great interest in the digitalization of justice and use of various online forms of redress and dispute resolution. The court modernization programme was introduced in 2016 in the UK (HM Courts & Tribunals Service 2018). Its basic premise was to provide new, user-friendly digital services and to improve efficiency of the justice process.

In 2016, Lord Briggs conducted a review of the courts in England and Wales with the aim of modernizing the system and making it more efficient. The review, commissioned by the Ministry of Justice, was undertaken in response to concerns about the cost and complexity of the court system and the slow pace of justice. One of Lord Briggs' conclusions about the modernization agenda was: 'The success of the Online Court will also be critically dependent upon digital assistance for all those challenged by the use of computers, and upon continuing improvement in public legal education.' The original vision for reform – to modernize and upgrade our justice system so that it works even better for everyone – remains true. But we must recognize that the world has changed since 2016 – and rapidly so – because of the COVID-19 pandemic that started in 2020 (HM Courts & Tribunals Service 2018; Sorabji 2021).

The pandemic happened at a time when the digitalization agenda was being carefully tested in those parts of the justice system in which judges allowed it to happen. The pilots were rolled out in stages (starting with Social Security Tribunals and followed by Tax Tribunals and Immigration Tribunals). A report by the Public Law Project (PLP) (2018), commissioned by the UK Administrative Justice Institute and the Nuffield Foundation, outlined the aims of the modernization agenda and quoted the Senior President of Tribunals who said that, unlike previous reforms, future reforms can no longer be predicated on the views of a single judge formed on the basis of anecdote or impression: 'reform must be based on proper research; robust and tested'. Some of the concerns about moving procedures online are that the best and established method of collecting evidence is in an oral hearing in a courtroom.

The report provides three reasons (PLP 2018: 22):

1. Other means of providing oral evidence may risk unfairness for appellants or reduce the ability of other parties to test such evidence.

2. The judicial task of collecting and evaluating facts and the credibility of the witness will depend not just on the appellant's oral evidence, but also on non-verbal forms of communication, e.g., how the evidence has been presented.

3. Giving live evidence at a hearing is subject to a degree of formality and supervision by the Tribunal. The procedure can be controlled to ensure that there is no misuse of the judicial process, aspects that will either be absent or reduced when video link is used.

These valid (pre-pandemic) fears had to be overcome, or reconsidered, when the justice system went online at a faster pace than planned when the pandemic started in 2020. We will come back to these stated fears and show how tribunals have dealt with these issues in Chapter 7. There is, however, an ongoing challenge which is to think about the 'traditional oral hearing' and what might be lost in video hearings (Hynes et al 2020: 7). Mulcahy states that there is 'no parallel call' for technology to replace physical presence at 'parliaments, weddings, christenings, bar mitzvahs or funerals' (2011: 178). Rowden (2018) argues that it is important to avoid nostalgia and idealism regarding the superior nature of hearings taking place with all participants physically present in open court. There are valid arguments and examples that show the benefits of online hearings (Hynes et al 2020), as well as valid arguments and examples that show the pitfalls of an online hearing (Sourdin et al 2020; Open Justice 2022). The following chapters will develop these in more detail, based on our empirical findings.

Our main themes in this book are access to (online) justice and vulnerability. These themes need to be understood in the context of the modernization agenda, which made promises in relation to all of them when it was being rolled out. In a speech in 2016 (Ryder 2016), the Senior President of Tribunals argued for the benefits an online dispute resolution system can have, and how it needs to be carefully and responsibly developed. He mentioned the importance of accessibility:

'Tribunals form an integral part of our country's justice system. They are and will continue to be an essential component of the rule of law; and must remain as accessible as possible. Accessibility is, however, not an unchanging construct. As society modernises, so must the institutions that serve it if they are not to degrade or fall into disuse.'

He goes on to mention access to justice:

'We should not forget that access to justice is an indivisible right: it is one that applies as much to defendants as it does to claimants. It is as important to ensure that meritorious claims are brought, and rights are vindicated, as to ensure that unmeritorious claims are resolved quickly and correctly so as to ensure the least interference with or disruption to the substantive rights of defendants. ... Citizens, whether litigants or not, are not supplicants coming to the high hand of judgement. They are rights bearers. And our justice system should be capable of ensuring that as such they are able to access those rights in an appropriate setting. Justice, and access to it, should lie at the heart of the community.'

In relation to vulnerable users, the modernization programme promised 'a justice system where many sizes fit all; not one size for all. A much simpler system of justice, with the judiciary at its heart, citizens empowered to access it, using innovation and digital tools to resolve these cases quickly, authoritatively, and efficiently'. This ambition remains a work in progress and, as we will show in this book, access to the online justice system during the pandemic exposed the flaws in the system. The COVID-19 pandemic was an important moment for justice systems. The data underpinning this book will speak to the future of the system by highlighting people's lived experiences and expectations of an online system. It shows the divide in perceptions and experiences of professionals and users when dealing with online hearings and raises pressing issues of vulnerability, marginalization and non-access.

Methodology

The project provides a novel and urgent empirical understanding of the ways in which people are accessing the system (and where they are not). We applied a mixed-methods approach to empirically understand access to, and trust in, administrative justice during the pandemic, to then draw lessons for a more efficient and fair justice system moving out of the pandemic. Our research methods were qualitative and quantitative, to best explore the population we were looking at. We accomplished this through vignette experiments with members of the general population (public panel) and interviews with those who administer the process (advice sector, ombuds, tribunal judges and case workers), those who use ombuds and tribunals, but we also interviewed marginalized groups who do not use the system. Table 0.1 provides an overview of the data collected. We obtained ethics clearance to conduct our planned research from the University of Westminster's ethics panel (ETH2223-0051).

Interviews

Overall, we conducted 58 in-depth semi-structured interviews (that is, the total of professional and user interviews; see Table 0.1). We conducted 40

Table 0.1: Overview of data collected

Data collected	Professionals n=40	Users (non–users) n=18	Members of the public n=480	Total N=538
Interviews	Judges (9) Ombuds (5) Advice providers (13) Other stakeholders (6) Institution staff (7)	SEND (6) Housing (12)	N/A	58
Experiments	N/A	N/A	480	480

in–depth interviews with professionals: nine judges; five ombuds; 13 advice providers; seven staff members at the institutions and six other stakeholders. Interview questions began by asking participants about their role and what their work entails. Participants were then asked about the most common issues they deal with in relation to housing/SEND. Next, questions revolved around the pandemic, getting participants to reflect on their experiences with people accessing their service during the pandemic and on any changes there have been to services because of COVID-19. Questions focused on methods of communication, benefits/downfalls of remote hearings, changes in user demographics, and reflections on what worked well/not so well in delivering remote justice during the pandemic and what could be changed to improve access for those further sidelined because of the pandemic. Finally, participants were asked whether institutions in the areas of housing and SEND have collaborated in any way to increase/improve access to justice for its users, and to reflect on whether a tribunals–ombuds partnership would be feasible.

We conducted 18 in–depth user interviews: six SEND users; and 12 housing users, including seven homeless people through The Connect (a charity supporting the homeless). Interview questions revolved around the eight steps we identified that users go through when seeking help (see Chapter 3). Interviewees were asked to share their stories, including questions around whether they had experienced any housing/SEND issues during the pandemic, at what point they became aware that there was a problem, and how they went about addressing that problem. Next interviewees were asked a series of questions on taking action, including whether they had tried to get support for their issues, how they looked for services, and whether they experienced any difficulties knowing how and where to look for help. Participants were also asked about the advice sector and any support or guidance they had received, before being asked to reflect on their experience of any intermediate processes involving their landlord/housing association in the case of housing or any organization involved (for example, local authority (LA), school or governing body) in the case of SEND. Those participants that had contacted a tribunal or ombuds were asked an additional set of questions revolving around how they went about accessing the justice system, which institution they approached, and how much time they spent trying to sort out their problem before approaching the institution, as well as any expectations they had. Next, they were asked to reflect on their experience of engaging with the institution, including what worked well, what barriers they faced and the extent to which they trusted the process. Finally, participants were asked what they thought could be done during and after the pandemic to improve users' capacity to obtain advice, support and redress.

Interviews were recorded and transcribed (with participants' permission) and were supplemented where relevant and practicable by our survey data.[3] We listened to the audio recordings and reflected on the survey responses as a team. We then iteratively winnowed the data and descriptions to focus on the most meaningful, relevant and revealing instances, stories and reports. The data that we decided best to represent the final set of themes were chosen collectively and are presented in the relevant chapters of the book.

Public panel survey including vignettes to examine trust in justice

We also conducted an online experimental study (see Chapter 6). The sample comprised 480 participants, who were roughly representative of the UK adult population. We used a text-based vignette describing a person going through a tribunal/ombuds process. We manipulated: (1) the fairness of the process (fair/unfair); (2) the location of the process (online/offline); and (3) the authority figure (judge/ombuds). We explored whether exposure to different tribunal/ombuds processes was accompanied by a concomitant loss of trust and legitimacy in the AJS, as well as damaging perceptions of process transparency and outcome fairness. Although the vignettes presented a hypothetical scenario, previous research has shown that varying behaviour through text-based vignettes can successfully shift participants' judgements of, for example, police legitimacy (for example, Silver 2020).

Recruitment of participants

The study was hosted on Qualtrics. Residents of England and Wales were recruited via the online crowdsourcing platform Prolific. In line with the Prolific recruitment protocols, participants received compensation for their time. We followed Chandler and Paolacci's (2017) advice on how to minimize participant fraud on Prolific: we set constraints so that participants could only take the survey once and included attention checks throughout the surveys. Participants were excluded if they got more than one attention check wrong.

Procedure and materials

Participants were presented with a short vignette about a person going through a tribunal/ombuds process. The study employed a 2 × 2 × 2 (fairness of process: fair/unfair × location of process: online/offline × authority figure: judge × ombuds) between-subjects design.

Participants were randomly allocated to one of eight conditions. They were presented with a vignette of the following:

1. a fair online tribunal process
2. an unfair online tribunal process
3. a fair offline tribunal process
4. an unfair offline tribunal process
5. a fair online ombuds process
6. an unfair online ombuds process
7. a fair offline ombuds process
8. an unfair offline ombuds process.

At the threshold, we should make clear that ombuds have always had primarily online processes (with the option for telephone interaction) so the 'offline' ombuds process depicted in vignettes 7 and 8 is hypothetical. In this experiment we had to create a comparison between online and offline scenarios to make reliable claims about online interactions. After reading the vignette, participants were asked a series of questions tapping into the quality of the process/outcome and the justice system more generally. Finally, they were provided with a full debrief.

Outline of the book

Part I: Situating Access to Justice, consists of two theoretical chapters that frame the book. In Chapter 1, Legal Needs and Access to Justice, we develop the argument for a holistic vision of access to justice (Creutzfeldt et al 2021). We expand Wrbka's (2014) definition of 'the concept of access to justice that embodies the ideal that everybody, regardless of his or her capabilities, should have the chance to enjoy the protection and enforcement of his or her rights by the use of law and the legal system' and argue that we need a broader definition. To date, access to justice is refined to a narrow 'legal justice' focus, involving access to legal assistance in the form of legal advice and access to resolution in the form of legal institutions. A more generous vision for access to justice is needed to include initial advice and help from non-legal support, social and community actors (for example, friends, family, advice sector, local council, specialist organizations (NGOs), schools, the internet) to be part of the delivery of access to justice. As part of this vision, we discuss the legal needs literature and propose a more generous approach to access to justice, reaching beyond legal confines. After that, we distinguish access to offline justice from access to online justice.[4] Then, we set out theoretical frameworks through which to understand (and measure) access to justice in our dataset; namely, legal consciousness and procedural justice.

Through the lens of procedural justice theory, Chapter 2, Trust in Administrative Justice, captures people's experiences of, and sensibilities towards, moving parts of the AJS online. Prior research has found that procedural justice and the trust and legitimacy it engenders helps to strengthen people's willingness to cooperate with the police, courts and other justice institutions, and to comply with their directives and the law in general. Yet, little is known about whether and how this process 'works' in an administrative justice context within which interactions are increasingly occurring primarily, or solely, online. This chapter will also explore the role of emotions in the encounters with the AJS, marked by strong asymmetric dependence and power. This chapter provides the theoretical foundation for the analysis and experimental vignettes in Chapter 6.

Part II: Pathways to Justice, is made up of three chapters.

In Chapter 3, Two Areas of Law in Context and the Help-Seeker Journey, we provide the context for the pathways to SEND and pathways to housing. The help-seeker journey follows the person with a problem and legal need through different stages of seeking help, finding advice and reaching an ombuds or tribunal to resolve their problem. For housing, we look at the advice sector, the Property Chamber and the Housing Ombudsman; and for SEND we look at the advice sector, the SEND Tribunal, the LGSCO and the PHSO. The emphasis on these institutions allows us to understand in some depth the effects of the pandemic, how such institutions managed to provide their services remotely, and what lessons can be learned for the AJS and the justice system more generally.

Chapter 4, Pathways Through the AJS: Housing, explores the pathways to redress available to people through mapping the ideal case help-seeker journeys for people with issues around housing to understand how access points have been compromised and which pathways to justice are difficult to negotiate or even blocked. The Housing Ombudsman and the Property Chamber provide redress for housing problems. In this chapter we will draw on interviews conducted with advice sector professionals, judges, case handlers and users to trace the help-seeker journey.

Chapter 5, Pathways Through the AJS: SEND, follows a similar structure to Chapter 4. It explores the ideal case help-seeker journeys for people with issues around SEND to understand how access has been compromised and which pathways to justice are difficult to negotiate or blocked. The LGSCO, the PHSO and the SEND Tribunal provide redress for SEND problems. Here too, we will draw on interviews conducted with advice sector professionals, judges, case handlers and users to present how the help-seeker journey unfolds.

Part III: Exploring Help-Seeker Journeys, is made up of three chapters. We briefly introduce and bring together the themes from our empirical data discussed in this part. Our aim is to situate digital journeys by discussing the challenges and opportunities of technology in access to justice and providing support and guidance for digital and legal needs. Three overarching themes emerged from our empirical data that are relevant to examining procedural justice, trust in justice and access to digital justice in both housing and SEND contexts.

1. *Advancements in technology and access to justice.* Advancements in technology have opened new opportunities for accessing justice, particularly for those who were previously excluded. However, there are still barriers to access to justice, such as lack of digital literacy and limited access to technology. To overcome these barriers, we need to ensure that technology is accessible and user-friendly for all, regardless of their digital capabilities.
2. *Face-to-face hearings and trust in the legal system.* Face-to-face hearings can increase trust in the legal system. However, in the digital age, we are increasingly moving towards online hearings. It is essential to ensure that online hearings are designed to promote trust and legitimacy in the legal system, for example, by ensuring that they are transparent and that users have access to information about the process.
3. *Ensuring inclusive justice.* Inclusive justice requires us to consider the needs of vulnerable populations and ensure that legal processes do not disproportionately impact them. Marginalized groups often experience unmet legal needs and negative perceptions of legal services. It is important to provide tailored support to address their specific needs and to ensure that they have access to justice.

In Chapter 6, Exploring the Role of Procedural Justice in Tribunals and Ombuds, we draw on data and findings produced by our online experimental study. We consider the idea that experiencing procedural justice during tribunals and ombuds hearings is important not only in shaping legitimacy, but also in shaping perceptions of outcome fairness, satisfaction and willingness to engage with the system in the future. We also assess whether the findings are different for online and offline proceedings.

Chapter 7, Access to Digital Justice, asks the central question: how accessible is online justice? This chapter explores how those who administer justice, those who provide advice and those who use the online justice system experience it. In doing so, we explore how the use of technology in the justice system is shaped by, and may reshape, people's orientations and sensibilities towards law and technology. We use our data, in this chapter, to explore how consciousness of how people think and feel about the law relates to their capability of acting upon it.

Chapter 8, Marginalized Groups and Unmet Legal Needs, explores how the pandemic has affected access to advice and redress for marginalized groups. Already marginalized communities are likely to be affected the most by the pandemic. Yet, we know relatively little about how members of these groups are accessing the justice system and what can be done during and after the pandemic to improve their capacity to obtain advice, support and redress. In addressing these questions, the book builds upon, and seeks to extend, existing work about marginalized groups that are alienated by the justice system and whose relationships to authority are characterized by a context of structural disempowerment. In sum, the digital age has brought both opportunities and challenges to the access to justice landscape. While advancements in technology have made justice more accessible to some, they have also created barriers for others, especially those who are digitally and legally excluded. Our empirical data show that there is a need for inclusive justice that addresses the unmet legal needs of marginalized groups, provides support and guidance for legal needs and promotes knowledge and awareness of tribunals and ombuds. Additionally, procedural justice plays a crucial role in establishing legitimacy in the digital age, and communication that considers both the form and content can have a significant impact on the emotional experiences of service users. It is essential to consider the unique vulnerabilities and capabilities of different groups and aim towards creating a justice system that serves everyone, including the digitally and legally abandoned.

Chapter 9, Conclusion, brings together the empirical findings of the project and critically assesses what we have learned from doing research with marginalized groups and how we might rethink the approaches to understanding access to justice. We offer a more nuanced understanding of people's digital journeys through bringing procedural justice to the concept of digital legal consciousness as well as three dimensions that came out of our data: digital, affective and compound. This wider perspective can help identify barriers to access and inform strategies to improve access to justice. Ultimately, a more fine-grained understanding of digital legal consciousness will require ongoing research and collaboration between legal practitioners, policymakers and technology experts.

Notes

[1]   We use *ombuds* to keep the term short and gender-neutral.

[2]   Project website: 'Delivering administrative justice after the pandemic', Nuffield Foundation. www.nuffieldfoundation.org/project/delivering-administrative-justice-after-the-pandemic; Creutzfeldt et al 2023.

[3]   We designed and distributed 11 surveys from June 2022 to November 2022: four user surveys (Housing Ombudsman, LGSCO, Property Chamber and SEND Tribunal); four case-handler surveys (Housing Ombudsman, PHSO, Property Chamber and SEND

Tribunal); two judicial and non-judicial panel members surveys (judges, SEND Tribunal, and judicial and non-judicial members of the Property Chamber); and one for the advice sector. However, despite our efforts to mitigate the low response rate, the final dataset had significant levels of missing data, rendering it unsuitable for our planned analyses. We were only able to produce descriptive statistics of the user sample available (N=40) and to run limited analyses using the more robust PHSO case-handler sample. Therefore, we could only use some of our open-ended survey responses to supplement our rich dataset.

[4] To explore the interaction with the digital justice space, theories of legal consciousness are brought to digital justice (more in Part III).

References

Adler, M. (2003) 'A socio-legal approach to administrative justice', *Law and Policy* 25(4): 323–352.

Adler, M. (2012) 'The rise and fall of administrative justice – a cautionary tale', *Socio-Legal Review* 8(2): 28–54.

Administrative Justice Council (2020) 'Ombudsman and Tribunals Familiarisation Workshop'. https://ajc-justice.co.uk/wp-content/uploads/2020/01/Ombudsman-and-Tribunals-Familiarisation-Workshop-Minutes-11-Oct.pdf

Buck, T., Kirkham, R. and Thompson, B. (2016) *The Ombudsman Enterprise and Administrative Justice*, London: Routledge.

Carl, S. (2012) 'Toward a definition and taxonomy of public sector ombudsmen', *Canadian Public Administration* 55(2): 203.

Carnwath, J. (2011) 'Tribunals and the courts – the UK Model', *Canadian Journal of Administrative Law and Practice* 24(1): 5.

Chandler, J.J. and Paolacci, G. (2017) 'Lie for a dime: when most prescreening responses are honest but most study participants are impostors', *Social Psychological and Personality Science* 8(5): 500–508.

Creutzfeldt, N. (2018) *Ombudsmen and ADR: A Comparative Study of Informal Justice in Europe*, London: Palgrave Macmillan.

Creutzfeldt, N. (2022) 'Ombuds and tribunals in a digital era: framing a digital legal consciousness', in Matthew Groves and Anita Stuhmcke (eds), *The Ombudsman in the Modern State*, Oxford: Hart, 141–164.

Creutzfeldt, N. and Sechi, D. (2021) 'Social welfare [law] advice provision during the pandemic in England and Wales: a conceptual framework' *Journal of Social Welfare and Family Law* 43(2): 152–174.

Creutzfeldt, N., Gill, C., Cornelis, M. and McPherson, R. (2021) *Access to Justice for Vulnerable and Energy-Poor Consumers: Just Energy?* Oxford: Hart Bloomsbury.

Creutzfeldt, N., Kyprianides, A., Bancroft, H., Bradford, B. and Jackson, J. (2023) 'How has the pandemic changed the way people access justice? Digitalisation and reform in the areas of housing and SEND', London: Nuffield Foundation. www.nuffieldfoundation.org/project/delivering-administrative-justice-after-the-pandemic

Drewry, G. (2009) 'The judicialisation of "Administrative" tribunals in the UK: from Hewart to Leggatt', *Transylvanian Review of Administrative Sciences* 5(28): 45–64.

Genn, H. (1993) 'Tribunals and informal justice', *Modern Law Review* 56(3): 393–411.

Good Things Foundation (2020) 'Building a digital nation'. www.goodt hingsfoundation.org/insights/building-a-digital-nation/

Groves, M. and Stuhmcke, A. (2022) *The Ombudsman in the Modern State*, Oxford: Hart.

HM Courts & Tribunals Service (2018). 'Guidance: The HMCTS Reform Programme'. www.gov.uk/guidance/the-hmcts-reform-programme

House of Commons (2023) 'Oversight of administrative justice'. www.par liament.uk/globalassets/documents/commons-committees/public-adm inistration/written-evidence-OAJ.pdf

Hynes, J., Gill, N. and Tomlinson, J. (2020) 'In defence of the hearing? Emerging geographies of publicness, materiality, access and communication in court hearings', *Geography Compass* 14(9): e12499.

JUSTICE (2020) 'JUSTICE response to HMCTS survey on conducting video hearings'. https://files.justice.org.uk/wp-content/uploads/2020/04/06165958/JUSTICE-Response-to-HMCTS-Survey-updated-images.pdf

Kirkham, R. (2016) 'The ombudsman, tribunals, and administrative justice section: a 2020 vision for the ombudsman sector', *Journal of Social Welfare and Family Law* 38(1): 103–114. DOI: 10.1080/09649069.2016.1145836

Kirkham, R. and Creutzfeldt, N. (2019) 'Reform of the administrative justice system: a plea for change and a research agenda', UK Administrative Justice Institute.

Kirwan, S. (2016) 'The UK Citizens Advice service and the plurality of actors and practices that shape "legal consciousness"', *Journal of Legal Pluralism and Unofficial Law* 48(3): 461–475.

Koch, I. and James, D. (2022) 'The state of the welfare state: advice, governance and care in settings of austerity', *Ethnos* 87(1): 1–21.

Leader, K. (2017) *'Fifteen stories: litigants in person in the civil justice system'* (Doctoral dissertation, London School of Economics and Political Science).

Legal Education Foundation (2019) 'Digital justice: HMCTS data strategy and delivering access to justice. Report and recommendations'. https://research.thelegaleducationfoundation.org/wp-content/uploads/2019/09/DigitalJusticeFINAL.pdf

Mayo, M., Koessl, G., Scott, M. and Slater, I. (2015) *Access to Justice for Disadvantaged Communities*, Bristol: Policy Press.

McKeever, G., Royal-Dawson, L., Kirk, E. and McCord, J. (2018) 'Litigants in person in Northern Ireland: barriers to legal participation – summary report'. https://papers.ssrn.com/sol3/papers.cfm?abstract_id=3523915

Mullen, T. (2010) 'A holistic approach to administrative justice', in M. Adler (ed), *Administrative Justice in Context*, Oxford: Hart, 383–420.

Mullen, T. (2016) 'Access to justice in administrative law and administrative justice', in E. Palmer, T. Cornford, Y. Marique and A. Guinchard (eds), *Access to Justice: Beyond the Policies and Politics of Austerity*, Oxford: Hart, 60–104.

Open Justice (2022) '"I am fearful for my daughter's life": serious medical treatment in a contentious case'. https://openjusticecourtofprotection.org/2022/08/23/i-am-fearful-for-my-daughters-life-serious-medical-treatment-in-a-contentious-case/

Public Law Project (2018) 'The digitalisation of tribunals: what we know and what we need to know', London: Public Law Project. https://publiclawproject.org.uk/content/uploads/2018/04/The-Digitalisation-of-Tribunals-for-website.pdf

Rowden, E. (2018) 'Distributed courts and legitimacy: what do we lose when we lose the courthouse?' *Law, Culture and the Humanities* 14(2): 263–281.

Ryan, M., Harker, L. and Rothera, S. (2020) *Remote Hearings in the Family Justice System: Reflections and Experiences*, London: Nuffield Family Justice Observatory.

Ryder, E. (2016) (Senior President of Tribunals) 'The modernization of access to justice in times of austerity', 5th Annual Ryder Lecture, University of Bolton. www.judiciary.gov.uk/wp-content/uploads/2016/03/20160303-ryder-lecture2.pdf

Ryder, E. (2019) 'Securing open justice', *Open Justice* 125–142.

Sechi, D. (2020) 'Digitisation and accessing justice in the community', London: Administrative Justice Council. https://ajc-justice.co.uk/wp-content/uploads/2020/04/Digitisation.pdf

Seneviratne, M. (2002) *Ombudsmen: Public Services and Administrative Justice*. Cambridge: Cambridge University Press.

Silver, J.R. (2020) 'Moral motives, police legitimacy and acceptance of force', *Policing: An International Journal* 43(5): 799–815.

Sorabji, J. (2021) 'Initial reflections on the potential effects of the Covid-19 pandemic on courts and judiciary of England and Wales', *IJCA* 12(2): 1.

Sourdin, T., Li, B. and McNamara, D.M. (2020) 'Court innovations and access to justice in times of crisis', *Health Policy and Technology* 9(4): 447–453.

Stebbings, C. (2006) *Legal Foundations of Tribunals in Nineteenth Century England*, Cambridge: Cambridge University Press.

Thomas, R. (2016) 'Current developments in UK tribunals: challenges for administrative justice'. https://ssrn.com/abstract=2766982 or http://dx.doi.org/10.2139/ssrn.2766982

Tomlinson, J. (2017a) 'A primer on the digitisation of administrative tribunals'. https://papers.ssrn.com/sol3/papers.cfm?abstract_id=3038090

Tomlinson, J. (2017b) 'The grammar of administrative justice values', *Journal of Social Welfare and Family Law* 39(4): 524–537.

Tomlinson, J., Sheridan, K. and Harkens, A. (2020) 'Judicial review evidence in the era of the digital state'. https://ssrn.com/abstract=3615312

Wrbka, S. (2014) *European Consumer Access to Justice Revisited*, Cambridge: Cambridge University Press.

PART I

Situating Access to Justice

1

Legal Needs and Access to Justice

Introduction

This chapter develops the argument for a holistic vision of access to justice (Creutzfeldt et al 2021). We expand upon Wrbka's (2014) definition of the concept of access to justice [that] embodies the ideal that everybody, regardless of their capabilities, should have the chance to enjoy the protection and enforcement of their rights by the use of law and the legal system, and thus argue that we need a broader definition. To date, access to justice has been refined down to a narrow 'legal justice' focus, involving access to legal assistance in the form of legal advice and access to resolution in the form of legal institutions. However, a more generous vision for access to justice is needed that includes initial advice and help from non-legal support, social and community actors (for example, friends, family, the advice sector, local councils, specialist non-governmental organizations, schools, the internet) as components of the delivery of access to justice. In light of this vision, we discuss the legal needs literature and propose a more generous approach to access to justice, reaching beyond legal confines. Following on from that, we distinguish access to *offline* justice from access to *online* justice,[1] and then set out theoretical frameworks through which to understand access to justice and analyse people's digital journeys in our dataset.

Legal needs and vulnerability

Legal needs surveys investigate the experience of justiciable problems from the perspective of those who face them (a 'bottom-up' perspective), rather than from that of justice professionals and institutions (a 'top-down' perspective). They seek to identify and explore the full range of responses to problems and, within this, all the various sources of help and expert institutions that are utilized in pursuing problem resolution. Such surveys provide a uniquely comprehensive overview that is impossible to achieve by other means.[2] Existing research based on legal needs surveys has demonstrated that those experiencing the greatest social and economic disadvantage and marginalization are often the least likely to take any action in response to a rights-based problem (Sandefur 2015b; Franklyn et al 2017). Similarly, the Legal Australia–Wide Survey found that some demographic configurations, including many disadvantaged groups, had heightened vulnerability to multiple legal problems (McDonald and Wei 2013).

The Organisation for Economic Co-operation and Development (OECD) has produced a guide on how to measure legal needs.[3] The guide provides a framework for understanding and measuring legal needs, as well as methodological guidance and model questions to capture three core components of effective access to justice. These are:

1. the nature and extent of unmet legal and justice needs;
2. the impact of unmet legal and justice needs on individuals, the community and the state; and
3. how specific models of legal assistance and dispute resolution are utilized to meet these needs.

In 2019, a legal needs study was conducted in England and Wales.[4] It was the first study in this jurisdiction to use OECD guidance on how to develop legal needs surveys. It includes measures of legal capability to profile the population by their experience and perceptions of the legal system.

Some of the key findings were (in a pre-pandemic study; Law Society 2019/20):

• six in 10 adults (64%) based in England and Wales had experienced a legal issue during the previous four years;
• 53% of people who had a contentious legal issue said they had experienced stress as a part of or as a result of it, and 33% had lost money;
• 55% of people with a legal issue had got professional help;
• 66% of those who had got professional help felt that the outcome was fair, compared to 54% who had either got non-professional help or no help at all;
• an estimated three in ten respondents had an unmet legal need for a contentious legal issue, where either they did not receive any help or had wanted more help to resolve their issue;
• 85% of people who had got help were satisfied with the service they received from their professional adviser, with people who received services from solicitors being most satisfied (90%);
• consumers with low legal confidence (that they could get a fair and positive outcome in different legal scenarios) had found it harder to find professional help and were more likely to get worse outcomes;

- 21% of people did not try to get help (for contentious or non-contentious issues) from a professional adviser because they assumed it would be too expensive; and
- 57% of those who did get professional help did not have to pay.

These figures clearly show that, as in other jurisdictions, people in England and Wales have compact legal needs and varying levels of trust in the process of seeking support and redress for their problems (Genn 1999; Pleasence et al 2004; Balmer et al 2006, 2010; Pleasence and Balmer 2014; Newman et al 2021). We will return to these findings and discuss them in relation to our data in Chapters 7 and 8.

As extensive surveys demonstrate, legal needs and vulnerability are closely connected for those people who, for a variety of reasons, are unable to easily identify their needs and therefore are not accessing advice, help and support. The much-debated term 'vulnerability' (Fineman 2013; Fineman and Fineman 2018; Newman et al 2021) is often used to better understand the complex nature of the different situations that people find themselves in. One of the challenges is to define vulnerability in such a way that provides us with helpful insights into the everyday experience of being vulnerable, but that also helps to build our theoretical framework around vulnerability. We can start to explore this through Fineman's (2017) theory of vulnerability. In her work, she attempts to understand the limitations of equality, and this struggle has resulted in the development of a legal paradigm that brings vulnerability and dependency, as well as social institutions and relationships, together into an analysis of state responsibility (2017: 134; see also Fineman 2008). Her theory is based on a descriptive account of the human condition as one of universal and continuous vulnerability. According to Fineman:

> the potential normative implications of the theory are found in the assertion that state policy and law should be responsive to human vulnerability. However, the call for a responsive state does not dictate the form responses should take, only that they reflect the reality of human vulnerability. Thus, this approach to law and policy allows for the adaptation of solutions appropriate to differing legal structures and political cultures. (Fineman 2017: 134)[5]

Consequently, a vulnerability approach to the *legal subject* starts from the premise that the concept needs to be rethought and made more representative of the actual human experience.

> It requires that we recognise the ways in which power and privilege are conferred through the operation of societal institutions, relationships and the creation of social identities, sometimes inequitably. Because

law should recognise, respond to, and, perhaps, redirect unjustified inequality, the critical issue must be whether the balance of power struck by law was warranted. (Fineman 2017: 142)

In contrast, McDowell's (2018: 104) approach takes vulnerability theory and focuses on the justice space, in particular, on access to justice. She suggests that: 'utilising these insights, and taking relative privilege, privacy, and autonomy into account, interventions into poor people's courts should seek not merely to provide access to existing legal systems, but also to mitigate the harm caused to low-income people using those systems, foster accountability, and develop meaningful alternatives.'

This requires a broad approach to providing access, including the provision of opportunities for people to develop the assets necessary for social, legal and political resilience and change. Attention to functional as well as problematic fragmentations in the state is one way to engage this project and create space for justice as well as for access.

In this book, through the lens of problems faced by individuals, we look at the space created for *justice* and for *access* in two distinct areas: housing and special educational needs and disabilities (SEND). These justice problems create the need for legal advice and legal assistance. As mentioned in the previous section, those people who most commonly face difficulties in these areas are not typically those who are even aware that they have a legal problem and that there is help available (Pleasence and Balmer 2014). Their legal needs are thus not being met. A legal need framing is helpful in understanding access to justice because we have to recognize why people use the justice system in the first place, and, consequently, the needs of an individual cannot be separated from that person's ability to engage with services. The relationship between individual needs and the ability to engage with services in the context of access to justice is not a straightforward one. It involves a series of constructivist feedback loops where the two factors are interdependent and influence each other in a continuous cycle. In a nutshell, individual needs and the ability to engage are inseparable because they are intimately connected.

For example, an individual's ability to access justice services is shaped by their level of literacy, language proficiency, financial resources, social support and mental and physical health. At the same time, their specific needs – for example, for legal representation or assistance with filling out forms – can influence their ability to engage with the relevant services. In this way, the two factors are constantly interacting with each other, and the relationship is not a static one. As an individual's needs change, so too can their ability to engage with the required services. Similarly, as their ability to engage with services improves, their needs may become more refined or complex. Thus, it is important to recognize at the outset how closely individual needs are interconnected with the ability to engage with services when it comes to accessing justice. This can inform the design of effective and responsive systems that are better suited to the needs of justice-seeking individuals.

Access to justice

By upholding the rule of law and ensuring access to justice, a society can strive for equality, fairness and accountability to foster trust in the legal system. We start this section with a discussion of the rule of law and how it relates to access to justice. The access to justice literature and the discussion around it is vast. We focus on the evolution from a very narrow law-focused definition of access to justice to a more inclusive and generous definition of access to justice, starting with introducing some of the literature framing the debate (Bailey et al 2013; Wrbka 2014).

Lord Neuberger (2013), in a lecture about the rule of law and access to justice, connected them firmly by arguing that access to justice is a rule of law requirement. Therefore, a practical consideration of the rule of law is access to the courts. He builds on this in another lecture in 2017 by saying:

> While access to law is important, access to legal advice and representation is equally important but more challenging. Access to legal advice and representation is of course a fundamental ingredient of the rule of law, and the rule of law together with democracy is one of the two principal columns on which a civilised modern society is based.

These assertions invite a further inquiry into the social and more inclusive understanding of the rule of law and of access to justice, beyond their purely legal confines. Barber (2004) explored two contrasting perspectives on the rule of law: the legalistic conception and the social conception. While the legalistic conception emphasizes formal legality and procedural fairness, the social conception recognizes the social and political dimensions of the rule of law. Both are important but the social conception provides a broader and more comprehensive understanding of the rule of law, considering the social impact and implications of the legal systems and processes. The social conception of the rule of law expands beyond the narrow focus on legal formalities and highlights the societal implications of the rule of law. It recognizes that the rule of law cannot be divorced from social, economic and political realities. The social conception emphasizes that the rule of law should promote justice, equality and social wellbeing. It considers the impact of legal institutions and processes on marginalized groups, power dynamics and the distribution of resources in society.

Lucy (2020) also explores the connection between access to justice and the rule of law. He develops connections between these concepts, based on a notion that society appears to take them for granted. One example of this is the Law Society of England and Wales (2019) stating that '[t]he rule of law underpins the very foundations of access to justice'. The entitlement to access to justice and the close ties to the rule of law are unpacked in detail in Lucy's paper. He concludes that the expansive nature of access to justice includes access to expertise and to legal knowledge itself (Law Society 2019: 8), beyond access to (legal) institutions. This chimes with our plea for understanding access to justice beyond its narrow legal understanding.

Adams-Prassl and Adams-Prassl (2020) used access to justice, embedded within the rule of law, to propose a framework for policy-level analysis that detects systemic unfairness. Their futility framework is firmly grounded in the principles of judicial review, but they argue that the structure and features will be equally relevant in the design and evaluation of online courts (Adams-Prassl and Adams-Prassl 2020: 35). What is interesting for our discussion is that they conclude that in determining the success of the court reform programme, access to justice is one of the most important metrics (MoJ 2016). The potential for digitalization to revolutionize access to justice is real: 'its streamlined efficiency can, when used properly, return effective dispute resolution in massive numbers of low value claims' (Endicott 2017). The authors acknowledge, however, that there are dangers of replicating existing patterns of exclusion or creating new barriers for the digitally illiterate (Rostatin 2019).

Access to justice, as a recent letter from the justice ministers of 16 countries to the United Nations Secretary-General put it, 'is best understood as the ability of people to resolve and prevent their justice problems, and to use justice as a platform to participate in their economies and societies'.[6] A 2022 literature review on access to justice (Weston 2022) acknowledges that access to justice is not only about courts and lawyers, but also about other informal and formal justice providers that address people's justice needs. This broader, more people-centred approach to access to justice is supported not only by academics and by policymakers: for example, one of the United Nation's sustainable development goals is to provide equal access to justice.[7] This calls for a broad view of access to justice, assessing the impacts of improved and expanded provision of justice services – both formal and informal – on economies, societies and the social contract. In relation to exacerbating disadvantage, as found in other studies mentioned in the previous section, members of society who are already disadvantaged face the highest barriers to accessing justice, and they are often hardest hit by its absence (Task Force on Justice 2019). A common driver of disadvantage is a lack of financial and legal capabilities. In other words, deficits in legal capability may be particularly important in understanding why individuals choose not to respond to justiciable problems (McDonald and People 2014).

The Access to Justice Foundation in the United Kingdom (UK) states on its website that:

> [T]wo thirds of the UK population don't know how to get legal advice and there are 14 million people living in poverty who can't afford it. A lack of access to justice leads to issues of poverty, homelessness, ill health, unemployment, broken households, and many other social and personal difficulties.[8]

Traditional conceptualizations of access to justice have focused on access to lawyers, legal expertise and legal institutions (Blankley 2020: 2021–2022). In contrast, a new approach that is now emerging is centred instead on justice itself. Poppe (2021: 781) finds that contemporary scholars integrate real-life difficulties into the definition of access to justice: namely, equality linked to the ability of individuals to achieve just resolutions to the problems they experience, regardless of whether they pursue formal legal action. To best achieve such a definition, the emphasis is on working out how groups most in need of assistance can address problems early to avoid escalation by focusing on people's capabilities. With the goal of equalizing individuals' ability to achieve just resolutions to legal problems regardless of whether lawyers or courts are involved, this approach pragmatically seeks to acknowledge individuals' disinclination to turn to law while nevertheless promoting their ability to achieve justice (Poppe 2021: 784). Also turning to the lived experience of access to justice, Farrow (2014) offers a lens through the voice of public opinion which declares access to justice as the most pressing justice issue today. Many legal problems still go unresolved (Rhode 2001): in the United States (US), 70–90% of the legal needs of citizens go unmet (Engler 2010: 40; citing Legal Services Corporation, 2009); in Canada, 65% of people are uncertain of their rights and do not know how to handle legal problems (Currie 2007). In the UK, findings of the 2019 survey of the *Legal Needs of Individuals in England and Wales* state that 64% of adults had experienced at least one legal issue in the previous four years, while 34% did not obtain help for their issues.[9] The results further find that having lower levels of legal confidence and low perceptions that justice is accessible are associated with being less likely to obtain professional help: of those with low *legal confidence*, 54% did not get professional help (versus 47% of those with *high confidence*). Similarly, half those who do not see justice as accessible did not get professional help (54%, versus 45% of those who perceive justice as accessible) (Law Society 2019/20: 4).

In light of this, one of the key aims of our study was to better understand both those who use the system and those who administer it. This inevitably leads us to also think about those who do not access the system. The rest of this book is therefore dedicated to the analysis of how the justice system is perceived not only by those who administer and access it, but also by those who do *not* access it. We situate our contributions in the growing literature on access to justice at the intersection of law and social justice. Recall, we do this by starting from the definition of access to justice offered by Wrbka (2014) as an ideal that everybody, regardless of their capabilities, should have the chance to enjoy the protection and enforcement of his or her rights by the use of law and the legal system. This understanding of access to justice needs to be stretched to reflect the true nature of (non) access to justice in its *lived form* (Albiston and Sandefur 2013). Therefore, we propose to add to the definition of access to justice the 'non-legal' components, informed by lived experiences of people who have different levels of vulnerabilities. We develop this broader and more user-focused understanding of access to justice in the following chapters, underpinned by our empirical data.

An important dimension of access to justice is the role of technology, which has been relevant in different ways to most people during the pandemic. Poppe discusses the technical advances that are changing law (2019: 185; see also McGinnis and Pearce 2014). These advances are reshaping the boundaries between clients and lawyers, with some futurists predicting a world in which lawyers become increasingly obsolete. Poppe uses an example of estate planning (will-preparation programmes) in the US to illustrate her argument about technology disrupting and enhancing access to justice. One of the points we want to draw out here is that of general and digital literacy. Despite the fact that many people are connected to the internet, this does not mean that they are able to access and understand all that is on offer (see Part III). For example, how do we distinguish good from bad sources of online information (Sandefur 2015a: 736)? The extent to which technology can address unmet legal needs, particularly for those less likely to benefit from technological interventions, remains unclear (Poppe 2019: 202). As mentioned in the previous section, many people do not seek advice for their legal problems as they do not identify them as such (Pleasence et al 2011: 11). And in the same vein, Poppe (2019) shows how technology is limited in the ways in which it can be used to help people with estate planning.[10] In relation to technology as legal assistance, she calls for grounding development in empirical realities (Poppe 2019: 212): 'Human frailties hinder the willingness and ability of many individuals to engage successfully with new technologies, while market forces shape the design and availability of technology in ways that may not address the needs of all.'

In the following section we introduce some examples of this tendency towards lack of engagement that we consider in light of the limitations and different needs of people accessing the online justice system.

Access to online justice

There are many obstacles that people face in accessing the justice system. For instance, existing challenges can be exacerbated by aspects of technology and online systems. The rapid switch to online provision of services is matched with a growing body of literature that attempts to understand the effects on users of online systems, especially vulnerable ones. One example is online video hearings which in many cases have now replaced in-person hearings.

As part of the court modernization programme in the UK (Lord Chancellor et al 2016),[11] the gradual transition to a more efficient online system has been carefully piloted. Academics (Rossner and McCurdy 2018) were asked to conduct a process evaluation of the user experiences of His Majesty's Courts & Tribunals Service (HMCTS) video-hearing pilot (for party-to-state hearings) in the Tax Tribunal (First–Tier Tribunal) in 2018. The pilot was testing the concept of video hearings. The data collection included interviews (with users, appellants, representatives and HMRC staff), observation of video and in-person hearings and interviews with judges managing these. A pre-selected handful of video hearings were tested, which received administrative support at the pre-hearing stage, and only when appellants and representatives were satisfied with this did they proceed to a video hearing. There were several themes that emerged, namely: suitability for a video hearing; user experience; technology difficulties and failures; authority of judicial proceedings; and judicial management and training (Rossner and McCurdy 2018: 1–3). One of the recommendations the authors made was that future developments need to include a strategy for addressing user vulnerabilities in video hearings or to identify a minimum standard of resilience (the antithesis of vulnerability) that one needs to meet in order to participate in a video hearing instead of an in-person hearing (Rossner and McCurdy 2018: 29). A minimum standard of resilience can be defined as ensuring that the user is provided with adequate support and accommodations to effectively participate in the hearing, including access to technology, interpretation services, and a safe and private space, in order to ensure a fair and equitable hearing process. The pilots were carefully planned, equipped with funds and technical support for all included, yet still exposed issues in the execution. The technical support was not intended to be available for parties after the pilot, which was bound to cause more problems. A follow-up report was published by the authors in 2020, with evidence collected before the pandemic (Rossner and McCurdy 2020). They issued 17 recommendations, drawn from key findings of the evaluation. These are grouped into recommendations to improve guidance (recommendations 1–3); recommendations about platform functionality (4–6); recommendations for improved products and associated services (7–9); and recommendations for future research

(10–17). In addition, recommendation 12 acknowledges a limitation of the pilot: namely, that data was not collected from vulnerable users who may need extra support during a court or tribunal appearance. Vulnerable groups in the population cannot be overlooked in the development of online systems.

A rapid consultation in 2020 on remote hearings in the family justice system found, among other things, that:

> Significant concerns were raised about the fairness of remote hearings in certain cases and circumstances, and there were some worrying descriptions of the way some cases had been conducted to date. These concerns chiefly related to cases where not having face-to-face contact made it difficult to read reactions and communicate in a humane and sensitive way, the difficulty of ensuring a party's full participation in a remote hearing, and issues of confidentiality and privacy. Specific concerns were commonly raised in relation to specific groups: such as parties in cases involving domestic abuse, parties with a disability or cognitive impairment or where an intermediary or interpreter is required. (Nuffield Foundation 2020: 1)

These concerns are still relevant today and readily apply to the tribunals in our study, as our data show in Chapter 7. However, some efforts to prepare people for a remote hearing have been made through the provision of online instructions,[12] information about joining the cloud video platform[13] and a video on how to prepare for a hearing.[14] These guides have been refined and improved over the past two years.

When considering access to online justice, other themes have been identified by Sanders (2021) who conducted a study about video hearings in Europe (before, during and after the pandemic) and found that, for the best use of remote hearings, two challenges need to be addressed: legal frameworks and technical solutions:

> Legal frameworks are needed which secure fair trial rights and provide appropriate access to the public and the media. Technical solutions are needed which are both user-friendly and protect sensitive data. Moreover, when traditional hearings are possible again, decisions must be made as to which hearings are suitable for video hearings and which conflicts might be solved more effectively with parties and judges present in one courtroom. (Sanders 2021: 2)

To us, this only scratches the surface of the challenges video hearings can bring with them, especially for those who are less confident and less able to operate effectively in the digital space (Chapters 7 and 8).

The example of video hearings is useful for thinking about both the benefits and obstacles to accessing an online justice system more widely. The aim of technology to assist, and in some cases to replace, humans in the justice system is an unstoppable evolution. The benefits are cost savings, accessibility (for those who are able enough), the promise of faster resolution and no time wasted with paper-based processes. Also, a digital system can enable people to work remotely (with appropriate equipment and a reliable internet connection) and reduce the amount of time and cost spent on travel. This is true for judges, case handlers and interpreters, as well as for appellants, as we will show in Chapter 7.

How is the digital transition experienced by the users of the system? To answer this question, we draw on theories of legal consciousness and procedural justice (Chapters 2 and 3) to structure our data. We analyse our empirical data in Part III.

Legal consciousness

How do people think about the law and authority in the digital space? As Ewick and Silby (1998) explain, legal consciousness is the process by which people make sense of their experiences by relying on legal categories and concepts. Legal consciousness research seeks to understand people's routine experiences and perceptions of law in everyday life (Sarat and Kearns 1995: 55). The focus is on subjective experiences, rather than on, for example, law and its effects in society (Cowan 2004). Legal consciousness allows us to study how people make sense of institutions and what they expect them to be doing: 'The study of legal consciousness traces the ways in which law is experienced and interpreted by specific individuals as they engage, avoid, or resist the law and legal meanings' (Silbey 2008).[15]

In a recent survey of the field of legal consciousness Halliday (2019) suggested that it had gained such interest because the concept has moved from its original theoretical orientation, within critical legal theory, to claim other areas. He lists these as four approaches: a critical approach; an interpretive approach; a comparative cultural approach; and a law–in–action approach (Halliday 2019: 3–13). The lens of legal consciousness lends itself to a wide application to study law in society. This study, if thinking of it through these four frames, can be situated in the *law-in-action approach* and adds a novel quantitative element. The law–in–action approach to legal consciousness, as Halliday argues, had developed out of law schools. It examines the social reality of law, the way that people experience law. The *law-in-action* approach is typically explored through qualitative interviews to understand its nuances. Our study, while following the methodological tradition of conducting interviews, adds a quantitative element and creates a new line of studying legal consciousness. Within this, we contribute to the

31

expanding literature on legal consciousness by both methodologically and theoretically developing the concept (see later on digital legal consciousness).

Our data cannot claim to be generalizable to reflect all people in all nuances and contexts of their interactions with the law. We agree with Nielsen (2000) that legal consciousness of ordinary citizens is not a single phenomenon, rather it is situated in relation to types of law, social hierarchies and the experiences of different groups through their encounters with law. Therefore, our focus is on those people who interact with a tribunal to resolve issues they face with housing or SEND problems. As outlined in Part II, these two pathways have moved online in the pandemic and are now slowly easing back into offering either in-person hearings or online hearings and, in some cases, a hybrid option. The digitalization agenda (see 'Introduction') has been fast-tracked by the pandemic and has provided us with a rich source of data to see what works well and what needs improvement. In our data we have uncovered a range of experiences with both online hearings and face-to-face hearings. Building on, and inspired by, the legal consciousness literature, in order to trace individual digital journeys, we use a taxonomy of attitudes of people who (dis)engage with the online justice system as our organizing framework (Creutzfeldt 2021).

Developing legal consciousness: users and non-users of the online justice system

One way to understand users and non-users of the SEND Tribunal and Property Chamber is to apply the lens of capabilities. We argue that people need legal and digital capabilities to access and navigate the online justice system, and build on emerging work in Australia and in the UK. Denvir et al (2021) explored the relationship between digital and legal capabilities through a quantitative analysis of data from the Community Perceptions of Law Survey in 2019; Creutzfeldt (2021) proposed a framework through which to understand legal and digital capabilities in relation to people's attitudes towards the online justice system.

Legal capability is understood as the 'elements of personal capability a person requires to be capable in the domain of the law and its institutions' (Pleasence et al 2014: 130). Legal capability, then, refers to an individual's ability to understand and navigate legal systems and processes, and to access legal information and resources to address legal issues that may arise in their lives. In practice, this means that individuals with legal capability are better equipped to protect their rights and interests, make informed decisions and engage with legal systems in a meaningful way. Some years on, Balmer and colleagues designed a framework and measures for legal capability (Pleasence and Balmer 2018, 2019; Balmer et al 2019). They expanded the measures for legal capability to include digital capability measures. As a result, in their quantitative analysis, they concluded that online systems do not provide a more accessible justice system. Further, that 'digital capability cannot be separated from legal capability, nor can either domain be understood without reference to the psychological factors that may operate to diminish capability in certain circumstances' (Balmer et al 2019: 54). In their analysis of cases, it was found that significant factors of personal circumstances are treated as immaterial in the decision-making process. In other words, the psycho-social needs of the individual (that might affect their capability to access the online system) are not considered. Moreover, Denvir et al (2021: 1) investigated how courts and tribunals dealt with digital and legal capabilities when handling a claim. One of their findings was that digital legal processes can undermine access to justice. Their data also showed that a lack of digital capability can impede fair outcomes for certain groups.

Digital capability is a term used to describe the skills and attitudes that equip individuals to live, learn, work and navigate in a digital society. The 2022 UK Digital Poverty Evidence Review (Allman 2022) found that:

> [A]round 11 million people in the UK lack the digital skills needed for everyday life, and 36% of the workforce lack Essential Digital Skills for Work. Moreover, only 74% of those who earn up to £13,500 per year have Essential Digital Skills for Life, compared to 95% of those who earn over £75,000. (Allman 2022: 10; see also Lloyds Bank 2021)

They go on to state that, despite policy developments that have responded to this evolving understanding of digital exclusion, they have also contributed to it. Examples of this include the efforts to modernize the courts and the introduction of digital-by-default services (Yates et al 2015). To start to fill this digital literacy gap, the Department of Education has, for example, funded initiatives such as the Digital Skills Partnerships.[16]

Intermediaries and local support have been mentioned often by our interviewees as crucial in getting them into, and guiding them through, their complaint journeys. We turn to these in more detail in Chapters 7 and 8. A combination of qualitative interview data and emerging quantitative measures will enable us to understand more about what motivates people to use or not use an online justice system. We begin doing this by understanding users, commencing with four distinct types, through the interplay between legal and digital capabilities in everyday lives (Table 1.1). These types allow us to start to explore the delicate relationship between digital and legal capabilities and the effect they have on experiencing justice online. In reality, there are no strict boundaries between each of the profiles; they are interconnected and, depending on the individuals and circumstances, they

Table 1.1: Digital legal consciousness

Digital capabilities

		+	−
Legal capabilities	+	Digitally literate and enabled Legally literate and enabled	Digitally agnostic Legally able
	−	Digitally assisted Legally unable	Digitally excluded and abandoned Legally excluded and abandoned

can change. They serve the purpose, however, to provide a framework to launch a discussion about how we can think about users in their digital journeys through the justice system in our dataset.

Type 1 – digitally and legally literate, enabled: people in this category are aware of the law, identify with the law, and choose to operate online. They are confident to use the justice system online and possess both digital and legal capabilities. Digitally literate citizens know how to access technology and are open to using technology to deal with all aspects of their lives (Carnegie Trust UK 2017).

Type 2 – digitally agnostic, legally able: people in this category are aware of the law, identify with the law, but choose not to use the internet for justice matters. Although they are able to use the online justice system, they draw a line and believe that justice needs to remain in an offline space. This group of people uses the internet with ease for other transactions as a routine part of their lives. In other words, they have legal capabilities but choose not to use their digital capabilities for engaging with the legal system online.

Type 3 – digitally assisted, legally unable: people in this category are aware of the law but have little identification with the law and are not able to digitally access the justice system themselves – they need assistance. Typically, this group of people will have regular engagement with the state welfare system and, possibly, designated advisers to assist them. They are not able to identify that the problems they are facing might be legal ones or where to get help. This group is on the margins of being digitally excluded, but if they manage to seek help then they will obtain assistance in legal and digital access. Although these people have low digital capabilities and low legal capabilities, with assistance they can access the digital justice system.

Type 4 – digitally and legally abandoned/excluded: people in this category have a limited legal awareness and are not able to access the digital justice system. They have turned their backs on the law, usually due to a lack of assistance or of their own capabilities. This group has lost trust in justice and thinks that it is just not for them. They suffer the most, as they do not have access to technology and do not know if their problems are legal problems or where they can go to seek help. These people are digitally excluded; they lack the skills and confidence to use online technology and usually do not have access to devices or stable internet connections. This is just one layer of social and economic problems that are also related to social exclusion.

In our study, we test how these basic types are represented in our data to understand better what the obstacles to accessing services are. As a starting point, the lens of capabilities is a helpful one to grasp the complex set of challenges people can face when engaging with the justice system online. What we do find, and this comes as no surprise, is that the lived experience of users of the system is far more complex than an abstract framework can capture. Especially interesting are the dimensions of vulnerabilities and emotions that many of our interviewees expressed – these can transcend the ideal types and change during a process. In other words, vulnerability can develop through the process (affecting any of the three types engaging with the system) and emotional responses can also evolve. These experiences have been described as side-effects of the process or the technology not working and all are new dimensions to consider when designing an online justice system.

To explore in more depth what people think about the law and legal processes, we add procedural justice to legal consciousness. As suggested by Young (2014), bringing procedural justice and legitimacy literatures to bear on legal consciousness may help to disaggregate people's beliefs about various types of legitimacy and understand how conditions, characteristics and experiences produce multidimensional orientations toward law. Tyler (2004) finds, in his research on the police, that people are more likely to support the police if they see police as legitimate authorities, and are more likely to see police as legitimate authorities if they believe the police act fairly. People may comply with authorities' rules because they believe that everyone else thinks the authority is legitimate. Here, citizens' perceptions of state actors define legitimacy as 'a property of an authority or institution that leads people to feel that that authority or institution is entitled to be deferred to and obeyed' (Sunshine and Tyler 2003).

Procedural justice, as we will discuss in Chapter 2, refers to the concept of fairness in the processes and procedures used to make decisions and resolve disputes. It is essential to ensure that citizens trust the authorities and view them as legitimate. When individuals perceive that their voices have been heard and that they have been treated fairly and impartially, they are more likely to comply with the outcome and accept it as just, even if they disagree with it. In the context of ombudsmen and tribunals, applying the principles of procedural justice can lead to greater trust and confidence in these institutions. For instance, ombudsmen and tribunals can ensure that their procedures are transparent, accessible and impartial. They can also provide individuals with an opportunity to voice their concerns and present evidence, and to have their cases reviewed by an independent third party.

By doing so, these institutions can demonstrate that they are committed to treating all individuals fairly and justly, which can enhance their legitimacy and credibility in the eyes of the public. Overall, by applying the principles of procedural justice, ombudsmen and tribunals can contribute to the development of a more just and trustworthy society. In addition, digital access requires procedural justice for legitimacy. Exclusion from the digital sphere and negative impressions of online justice can mean that fairness is perceived differently. For digital legal consciousness to emerge, we need to understand how access to justice can be made fair online. We go into more detail on this in Chapter 2 where we explore procedural justice.

Conclusion

In conclusion, this chapter argues for a wider, more inclusive vision of access to justice that goes beyond the narrow focus on legal justice. The definition of access to justice needs to be understood to include initial advice and help from non-legal support, social and community actors. The legal needs literature is discussed, and a more generous approach is proposed to reach beyond legal confines. Additionally, the chapter distinguishes access to offline justice from access to online justice, and theoretical frameworks are set out to understand access to justice in our own dataset. Extending legal consciousness literature, a framework of capabilities (digital and legal) is suggested, with four types as a starting point, to unpack the complex nature of a multidimensional understanding of access to digital justice (Part III). It is evident that a more inclusive vision of access to justice is necessary to provide protection and enforcement of the rights of everyone, regardless of their capabilities. Access to justice should not be limited to legal assistance and resolution, but should also include the involvement of social and community actors in the delivery of access to justice. Chapter 2 discusses trust in administrative justice, another theoretical lens through which to understand our data.

Notes

[1] To explore the interaction with the digital justice space, theories of legal consciousness are brought to digital justice (see further in Chapters 7 and 8).

[2] The OECD guide sets out a framework for the design, implementation and analysis of legal needs surveys. It provides tools in a modular way so that they can be applied in different types of surveys. A global legal needs indicator is proposed, to be able to better understand the different needs and vulnerabilities of the population.

[3] See 'Legal needs surveys and access to justice'. www.oecd.org/gov/legal-needs-surveys-and-access-to-justice-g2g9a36c-en.htm

[4] The work represents the biggest legal needs survey ever conducted in England and Wales, based on data collected online between February and March 2019 from 28,663 people. This sample is broadly representative of the population of England and Wales and covers 34 different legal issues (Law Society 2019/20).

[5] For more on vulnerability theory, see Rich (2018).

[6] Pathfinders for Peaceful, Just and Inclusive Societies (14 April 2021), 'Ministerial Meeting: building peaceful and inclusive societies through justice for all'. www.justice. sdg16.plus/ministerial

[7] See United Nations. https://sdgs.un.org/goals/goal16

[8] See Access to Justice Foundation. https://atjf.org.uk

[9] See 'Largest ever legal needs survey in England and Wales' (27 January 2020). https://leg alservicesboard.org.uk/news/largest-ever-legal-needs-survey-in-england-and-wales

[10] These are: internet access, web use and digital literacy (201); financial cost (202); psychic cost and avoidance (202); desire for legal expertise (203); boundary problems (204); and validity and enforceability (208).

[11] Other examples of reform across the justice system include an online divorce application service, a paperless system of sentencing for fare evaders and fraudulent ticket holders at a magistrate's court, and a service to lodge tax appeals online. For more on the reform programme, see HMCTS Reform Programme. www.gov.uk/government/news/hmcts-reform-programme

[12] See 'Guidance: HMCTS Video Hearings Service'. www.gov.uk/guidance/hmcts-video-hearings-service-guidance-for-joining-a-hearing

[13] See 'How to join Cloud Video Platform (CVP) for a video hearing'. www.gov.uk/gov ernment/publications/how-to-join-a-cloud-video-platform-cvp-hearing/how-to-join-cloud-video-platform-cvp-for-a-video-hearing

[14] HMCTS, 'Preparing for your video hearing'. https://youtu.be/0gzA1JIHhHA

[15] For more on this subject, see also Ewick and Silbey (1998); Silbey (2005, 2015); Ewick (2015).

[16] Department for Digital, Culture, Media and Sport, 'Guidance: Digital Skills Partnership', 19 October 2018. www.gov.uk/guidance/digital-skills-partnership

References

Adams-Prassl, A. and Adams-Prassl, J. (2020) 'Systemic unfairness, access to justice and futility: a framework', *Oxford Journal of Legal Studies* 40(3): 561–590.

Albiston, C.R. and Sandefur, R.L. (2013) 'Expanding the empirical study of access to justice', *Wisconsin Law Review* 2013(1): 101–120.

Allman, K. (2022) *UK Digital Poverty Evidence Review 2022*, Ascot: Digital Poverty Alliance.

Bailey, J., Burkell, J. and Reynolds, G. (2013) 'Access to justice for all: towards an "expansive vision" of justice and technology', *Windsor Yearbook of Access to Justice* 31(2): 181–207.

Balmer, N.J. and Pleasence, P. with Hagland, T. and McRae, C. (2019) *Law ... What Is It Good For? How People See the Law, Lawyers and Courts in Australia*, Melbourne: Victoria Law Foundation.

Balmer, N., Pleasence, P., Buck, A. and Walker, H.C. (2006) 'Worried sick: the experience of debt problems and their relationship with health, illness and disability', *Social Policy and Society* 5(1): 39–51.

Balmer, N.J., Buck, A., Patel, A., Denvir, C. and Pleasence, P. (2010) *Knowledge, Capability and the Experience of Rights Problems*, London: PLEnet.

Barber, N.W. (2004) 'Must legalistic conceptions of the rule of law have a social dimension?', *Ratio Juris* 17: 474–488.

Blankley, K.M. (2020) 'Online resources and family cases: access to justice in implementation of a plan', *Fordham Law Review* 88: 2121.

Carnegie Trust UK (2017) '#NotWithoutMe: A digital world for all?' www.carnegieuktrust.org.uk/publications/digitalworld/

Cowan, D. (2004) 'Legal consciousness: some observations', *Modern Law Review* 67(6): 928–958.

Creutzfeldt, N. (2021) 'Towards a digital legal consciousness?' *European Journal of Law and Technology* 12(3).

Creutzfeldt, N., Gill, C., Cornelis, M. and McPherson, R. (2021) *Access to Justice for Vulnerable and Energy-Poor Consumers: Just Energy?* London: Bloomsbury.

Currie, A. (2007) *The Legal Problems of Everyday Life: The Nature, Extent and Consequences of Justiciable Problems Experienced by Canadians*, Ottawa: Department of Justice Canada.

Denvir, C., Sutherland, C., Selvarajah, A.D., Balmer, N. and Pleasence, P. (2021) *Access to Online Courts: Exploring the Relationship between Legal and Digital Capability* (2 May 2021). https://ssrn.com/abstract=3838153 or http://dx.doi.org/10.2139/ssrn.3838153

Endicott, T. (2017) 'The rule of law and online dispute resolution', in Alessia Facheci, Timothy Endicott and Antonio Estella de Noriega (eds), *Online dispute resolution: virtud cívica digital, democracia y derecho*, Madrid: CEU Ediciones, 21–36.

Engler, Russell (2010) 'Connecting self-representation to civil Gideon: what existing data reveal about when counsel is most needed', *Fordham Urban Law Journal* 37(1).

Ewick, P. (2015) 'Law and everyday life', in J. Wright (ed), *International Encyclopaedia of the Social and Behavioural Sciences* (vol 13, 2nd edn), New York: Elsevier, 726–733.

Ewick, P. and Silbey, S.S. (1998) *The Common Place of Law: Stories from Everyday Life*, Chicago: University of Chicago Press.

Farrow, T. (2014) 'What is access to justice?' *Osgoode Hall Law Journal* 51(3): 957–988.

Fineman, M. (2008) 'The vulnerable subject: anchoring equality in the human condition', *Yale Journal of Law and Feminism* 20: 1.

Fineman, M. (2013) 'Equality, autonomy, and the vulnerable subject in law and politics', in M. Fineman and A. Grear (eds), *Vulnerability: Reflections on a New Ethical Foundation for Law and Politics*, Abingdon: Routledge, 13–28.

Fineman, M. (2017) 'Vulnerability and inevitable inequality', *Oslo Law Review* 4(3): 133–149.

Fineman, M. and Fineman, J.W. (eds) (2018) *Vulnerability and the Legal Organisation of Work*, New York: Routledge.

Franklyn, R., Budd, T. and Verill, R. (2017) *Findings from the Legal Problem and Resolution Survey 2014–15*, London: MoJ Analytical Services.

Genn, H. (1999) *Pathways to Justice: What People Do and Think about Going to Law*, London: Bloomsbury.

Halliday, S. (2019) 'After hegemony? The varieties of legal consciousness research'. *Social & Legal Studies* 28(6): 859–878. https://doi.org/10.1177/0964663919869739

Law Society (2019/20) 'Legal needs of individuals in England and Wales: summary report', London: Law Society/Legal Services Board. https://prdsitecore93.azureedge.net/-/media/files/topics/research/legalneedsofindividualssummaryreportjanuary2020.pdf?rev=67d5be937b0c41498234fb4a540a175a&hash=A59A40596C6A1C69BA08C366442E199C

Law Society (2019) 'Policy campaigns: our vision for law and justice'. www.lawsociety.org.uk/policycampaigns/articles/our-vision-for-law-and-justice-2019

Legal Services Corporation (2009) *Documenting the Justice Gap in America: The Current Unmet Civil Legal Needs of Low-Income Americans*, Washington, DC: Legal Services Corporation.

Lloyds Bank (2021) 'Essential Digital Skills Report 2021: third edition – benchmarking the essential digital skills of the UK', London: Lloyds Bank. www.lloydsbank.com/assets/media/pdfs/banking_with_us/whats-happening/210923-lb-essential-digital-skills-2021-report.pdf

Lord Chancellor, the Lord Chief Justice of England and Wales, and the Senior President of Tribunals (2016) *Transforming our Justice System*, London: Ministry of Justice.

Lucy, W. (2020) 'Access to justice and the rule of law', *Oxford Journal of Legal Studies* 40(2): 377–402.

McDonald, H.M. and People, J. (2014) 'Legal capability and inaction for legal problems: knowledge, stress and cost', *Updating Justice* 41, Law and Justice Foundation of New South Wales. http://www.lawfoundation.net.au/ljf/site/templates/UpdatingJustice/$file/UJ_41_Legal_capability_and_inaction_for_legal_problems_FINAL.pdf

McDonald, H.M. and Wei, Z. (2013) 'Concentrating disadvantage: a working paper on heightened vulnerability to multiple legal problems', *Updating Justice* 24: 1–5, Law and Justice Foundation of New South Wales. https://citeseerx.ist.psu.edu/viewdoc/download?doi=10.1.1.729.1761&rep=rep1&type=pdf

McDowell, E.L. (2018) 'Vulnerability, access to justice, and the fragmented state', *Michigan Journal of Race and Law* 23(1&2): 51–104.

McGinnis, J.O. and Pearce, R.G. (2014) 'The great disruption: how machine intelligence will transform the role of lawyers in the delivery of legal services', *Fordham Law Review* 82: 3041.

MoJ (2016) Lord Chancellor, Lord Chief Justice, and Senior President of Tribunals, *Transforming our Justice System* (Ministry of Justice) 5.

Neuberger (2013) Justice in an Age of Austerity: Tom Sargant Memorial Lecture 2013. www.supremecourt.uk/docs/speech-131015.pdf

Neuberger (2017) Access to Justice Welcome address to Australian Bar Association Biennial Conference. www.supremecourt.uk/docs/speech-170703.pdf

Newman, D., Mant, J. and Gordon, F. (2021) 'Vulnerability, legal need and technology in England and Wales', *International Journal of Discrimination and the Law* 21(3): 230–253. https://journals.sagepub.com/doi/full/10.1177/13582291211031375

Nielsen, L.B. (2000) 'Situating legal consciousness: experiences and attitudes of ordinary citizens about law and street harassment', *Law & Society Review* 34: 1055.

Nuffield Foundation (2020) 'Remote hearings in the family justice system: a rapid consultation', London: Nuffield Family Justice Observatory.

Pleasence, P. and Balmer, N.J. (2014) 'How people resolve "legal" problems', London: Legal Services Board.

Pleasence, P. and Balmer, N.J. (2018) *Legal Confidence and Attitudes to Law: Developing Standardised Measures of Legal Capability*, Cambridge: PPSR.

Pleasence, P. and Balmer, N.J. (2019) 'Development of a general legal confidence scale: a first implementation of the Rasch measurement model in empirical legal studies', *Journal of Empirical Legal Studies* 16(1): 143–174.

Pleasence, P., Balmer, N.J., Buck, A., O'Grady, A. and Genn, H. (2004) 'Multiple justiciable problems: common clusters and their social and demographic indicators', *Journal of Empirical Legal Studies* 1(2): 301–329.

Pleasence, P., Balmer, N.J. and Reimers, S. (2011) 'What really drives advice seeking behaviour? Looking beyond the subject of legal disputes', *Oñati Socio-Legal Series* 1(6): 1. http://opo.iisj.net/index.php/osls/article/viewFile/56/227

Pleasence, P., Coumarelos, C., Forell, S. and McDonald, H.M. (2014) *Reshaping Legal Assistance Services: Building on the Evidence Base: A Discussion Paper,* Sydney: Law and Justice Foundation of New South Wales.

Poppe, E.S. Taylor (2019) 'The future is complicated: AI, apps and access to justice', *Oklahoma Law Review* 72(1): 185–212.

Poppe, E.S. Taylor (2021) 'Institutional design for access to justice', *University of California Irvine Law Review* 11(3): 781–810.

Rhode, D.L. (2001) 'Access to justice', *Fordham Law Review* 69(5): 1785–1820.

Rich, P. (2018) 'What can we learn from vulnerability theory?', ScholarWorks@BGSU, Honors Projects 352, 30 April.

Rossner, M. and McCurdy, M. (2018) 'Implementing video hearings (party-to-state): a process evaluation', London: Ministry of Justice.

Rossner, M. and McCurdy, M. (2020) 'Video hearings process evaluation (phase 2)', London: HMCTS. https://openresearch-repository.anu.edu.au/bitstream/1885/261879/1/01_Rossner_Video_Hearings_Process_2020.pdf Final Report

Rostain, T. (2019) 'Techno-optimism & access to the legal system', *Daedalus* 148(1): 93–97.

Sandefur, R.L. (2015a) 'Bridging the gap: rethinking outreach for greater access to justice', *University of Arkansas Little Rock Law Review* 37(4): 721–740.

Sandefur, R.L. (2015b) 'What we know and need to know about the legal needs of the public', *South Carolina Law Review* 67: 443–460.

Sanders, A. (2021) 'Video-hearings in Europe before, during and after the Covid-19 pandemic', *International Journal for Court Administration* 12(2): 1.

Sarat, A. and Kearns, T. (1995) 'Beyond the great divide: forms of legal scholarship and everyday life', in A. Sarat and T. Kearns (eds), *Law in Everyday Life*, Ann Arbor: University of Michigan Press, 21–62.

Silbey, S.S. (2005) 'After legal consciousness', *Annual Review of Law and Social Science* 1(1): 323–368.

Silbey, S.S. (2008) 'Legal consciousness', in *New Oxford Companion to Law*, Oxford: Oxford University Press. https://web.mit.edu/~ssilbey/www/pdf/Legal_consciousness.pdf

Silbey, S.S. (2015) 'Legal culture and legal consciousness', in J. Wright (ed), *International Encyclopaedia of the Social and Behavioural Sciences* (vol 13, 2nd edn), New York: Elsevier, 468–473.

Sunshine, J. and Tyler, T.R. (2003) 'The role of procedural justice and legitimacy in shaping public support for policing', *Law & Society Review* 37(3): 513–548.

Task Force on Justice (2019) 'Justice for All – final report', New York: Center on International Cooperation. www.justice.sdg16.plus/report-old2022

Tyler, T.R. (2004) 'Enhancing police legitimacy'. *The Annals of the American Academy of Political and Social Science* 593(1): 84–99.

Weston, M. (2022) *The Benefits of Access to Justice for Economies, Societies, and the Social Contract: A Literature Review*, New York: Pathfinders for Peaceful, Just and Inclusive Societies.

Wrbka, S. (2014) *European Consumer Access to Justice Revisited*, Cambridge: Cambridge University Press.

Yates, S.J., Kirby, J. and Lockley, E. (2015). "'Digital-by-default": reinforcing exclusion through technology', in F. Foster, A. Brunton, C. Deeming and T. Haux (eds), *In Defence of Welfare*, Bristol: Policy Press, 158–161.

Young, K.M. (2014) 'Legal consciousness in the Hawaiian cockfight', *Law & Society Review* 48: 499–530. https://doi.org/10.1111/lasr.12094

2

Trust in Administrative Justice

Introduction

Questions of consent, cooperation and compliance have confronted the police and courts for many years. The ability to generate trust and command legitimacy is central to the functioning of legal institutions, their provision of services and delivery of outcomes (Tyler and Nobo 2023). On the one hand, people are more likely to use legal services, come forward with information, report crimes to the police and give evidence in court when they believe that justice institutions are trustworthy, appropriate and entitled to be obeyed. On the other hand, authorities that are seen as untrustworthy and illegitimate find it difficult to garner cooperation, gain acceptance of their decisions and enforce rules.

Scholars have produced a rich amount of academic research on the foundations, predictors and potential outcomes of legitimacy and 'trust in justice' (Tyler 2006b; Tyler and Huo 2002), especially in the context of the police, although this work is increasingly being applied in administrative justice and alternative dispute resolution (ADR) settings (Creutzfeldt and Bradford 2016; Bradford et al 2023; Hough 2020). In addition to highlighting 'downstream' implications for motivating and enabling engagement between two parties, this work has accumulated an increasing amount of evidence that trust and legitimacy are partly a product of people's experiences, judgements of and responses to their direct and indirect contacts with the police, courts and other legal entities. Justice system encounters represent 'teachable moments', whereby people update their attitudes towards the trustworthiness and legitimacy of legal authorities (can officials be trusted to do what they are supposed to do? Does the institution have the right to power and authority to govern?) based on how they perceive the official to behave, specifically the fairness exercised by legal authorities (Geller et al 2014).

Procedural justice scholars have focused on how courts, tribunals, police and other authorities treat people and make decisions (Mazerolle et al 2013) and have considered some of the potential implications for compliance (Tyler

2006b), cooperation (Tyler and Fagan 2008), empowerment (Sunshine and Tyler 2003) and community cohesion (Tyler and Jackson 2014) arising from the experience of procedural justice and the trust and legitimacy it does or does not generate. Research shows that people's attitudes and behaviours are shaped by the fairness of the processes that authorities offer them. While it has been suggested that improving upon the objective performance of legal authorities may not be enough in and of itself to enhance perceptions of fairness and legitimacy (Nagin and Telep 2017, 2020), there is strong evidence that the *subjective* impression of fair or unfair process influences how people perceive an institution and how they act in relation to it (Tyler 2003, 2017; Bolger and Walters 2019; Walters and Bolger 2019). It is therefore important to understand people's *perceptions* of the fairness of interpersonal treatment and decision-making.

In this chapter, we consider the role that procedural justice plays in public trust and institutional legitimacy. Because the majority of research has addressed the police, the literature we review mostly has that focus. We do, however, draw out prior work on the civil courts and ADR and consider applicability to administrative justice. We begin by defining two complex concepts: legitimacy and trust. We then consider the nature of procedural justice and its psychological significance in terms of its relational qualities and its effects on uncertainty management and emotions. We finish by framing the qualitative and quantitative work found in Part III.

What is legitimacy?

The concept of *legitimacy* has been approached in a variety of different ways (Tyler 2006a; Bottoms and Tankebe 2012; Hamm et al 2022). We frame legitimacy as a psychological orientation towards an authority that can be defined as 'the right to rule and the recognition by the ruled of that right' (Jackson et al 2012: 1015). It is this recognition – and the motivations and actions that flow from it – that constitutes legitimacy 'on the ground', as it exists in the relationship between an institution and those it governs. Within policing and criminal justice contexts, legitimate authorities are believed (by the governed) to have the right to issue commands, enforce laws and make decisions. A legitimate authority is one that people accept and obey because they believe it is right and proper to do so – not, for example, because they fear the consequences of failing to accept and obey.

When it comes to civil and administrative courts and ADR, a definition of legitimacy that focuses in part on normatively grounded deference and 'duty to obey' might be less intuitive than it seems in the context of policing or the criminal courts. But it is desirable that people should be willing to accept the decisions and recommendations delivered by tribunals and ombuds services, both for the smooth running of those services and in a wider, normative sense. Legal institutions of all kinds should behave in such ways that they are able to secure the willing acceptance and compliance of those who access them. Moreover, legitimacy judgements do not only reference and rely on notions of obedience, responsibility and deference. The recognition of the (rightful) authority to make decisions and expect compliance is rooted in assessments of the normative appropriateness of that legal institution. A sense of collective agreement on commonly shared norms, values and goals and, in particular, on how to achieve them is also of central importance.

On this account, legitimacy depends upon – and is in part defined by – people's sense of 'normative alignment' with institutions (Trinkner et al 2018). It reflects the extent to which they believe they share salient aims and values, and that the individuals and organizations that represent the institution are seen to enact and work according to normative expectations regarding the appropriate use of power. Perceptions of a shared sense of right and wrong and the belief that institutional actors are 'trying to do the right thing' when wielding their authority lead people to accept the institution's implicit claim that it is justified in expecting their support and consent. It is this *normative alignment* between police and public that gives the former the right to rule in the eyes of the latter, thereby generating willing self-regulation.

What is trust?

If legitimacy is about authority relations and approval of power, trust is about social coordination and acceptance of risk. A fair amount of work has cohered around the idea, first popularized by Mayer and colleagues (Mayer et al 1995), that trust is a willingness to accept vulnerability (a psychological state) that is driven by an individual's (a) perceptions of trustworthiness and (b) propensity to trust. On this account, trust involves a willingness to take *risk* – a willingness to be vulnerable to the agency and motives of another and to take the 'leap of faith' that the psychological state of trust represents (Möllering 2001). Thus framed, trust in the police manifests as a willingness to be vulnerable to police behaviour under inherent conditions of uncertainty, rooted in the perceived trustworthiness of the police to be effective, fair, and so forth. In addition to the role that legitimacy plays, trust has been identified as a key mechanism through which perceptions of officer and organizational performance and behaviour motivate people to proactively cooperate with the police (Hamm et al 2017).

Risk is a necessary feature of trust relations; the trustor cannot know for sure that the trustee will fulfil their promise to help them, solve their problem, safeguard their interests or undertake whatever action or actions the situation and the relationship demand and entail. People are more likely to cooperate with an individual or organization – who may of course not only fail to fulfil their duties but actively harm the trustor or their interests – when they make a judgement that the entity is trustworthy (and when they have a general propensity to trust) because this makes them more willing to take the risks cooperation implies.

Trust can thus be defined as a process that is founded in the propensity to trust and perceptions of trustworthiness and crystallized in a willingness to be vulnerable under conditions of risk. All are central to understanding how trust is formed and reproduced and what it does. From an organizational perspective, however, it may make sense to pay most attention to trustworthiness – to the actions and behaviours that lead people to feel the organization can be trusted – not least because this is where lessons about how to maintain and enhance trust relations can most readily be learned.

Trustworthiness judgements are commonly defined as the trustor's evaluations and expectations of the competency and good intentions of the trustee, to do what they are trusted to do (in the current context, for instance, for ombuds and tribunal officials to be fair when interacting with citizens and making decisions) (Hardin 2002; PytlikZillig and Kimbrough 2016; Hamm et al 2017). The precise content of such judgements – what it is that makes an entity trustworthy – is likely to vary across contexts, but most accounts stress the importance of technical ability, benevolence and integrity (Mayer et al 1995; Hamm et al 2017: 7). Additional factors might include reliability, openness and transparency, and identification between the two parties. Most if not all of these are actionable quantities for an organization; things it can hope to both measure or otherwise assess and change through policy and practice (unlike the propensity to trust, which is characteristic of people).

In justice contexts, trust and legitimacy are likely to be closely inter-linked (Jackson and Gau 2016). At the simplest level, it can be argued that perceptions of trustworthiness – of the behaviour, actions of, and outcomes secured by a legal authority – indicate its normative appropriateness: the things we think an authority *should do or achieve*, and our assessments of whether it actually does and/or achieves these things, both delineate our judgements that it is trustworthy (or not) and indicate that it is acting appropriately and in line with our contextual concept of right and wrong (or not). For a variety of reasons – some of which we expand on in the next sections – people believe that legal authorities such as tribunals and ombuds ought to act in certain ways (that is, should be trusted to act in those ways) and the perception that they do in fact act in those ways indicates that they are trustworthy actors that have the right to power and authority to govern.

What is procedural justice?

Assessments of procedural justice are central to perceptions of legitimacy and trust, not least because they encompass several characteristics listed in the previous section, including benevolence, integrity, reliability, transparency, neutrality, a lack of bias, and so on. Four key elements of procedural justice have been identified: offering participation or voice; making neutral, unbiased decisions; treating people with dignity and respect; and displaying trustworthy motives (Tyler and Fagan 2008; Tyler and Blader 2000). These are sometimes corralled under the broad headings of 'quality of decision-making' and the 'quality of treatment' (Tyler and Blader 2000; Tyler and Huo 2002). Quality of decision-making refers primarily to openness, consistency, neutrality and a lack of bias (and therefore, in general, the ability to make the right decision), while quality of interaction relates primarily to issues of respect, dignity and voice (and therefore, benevolence and good intentions). A further distinction is drawn between formal and informal levels (that is, those relating to formal rules and those relating to the behaviour of individual office holders (Tyler and Blader 2003)).

The core argument of procedural justice theory is that people who interact with legal and other authorities care deeply about the processes through which decisions are reached at least as much, and perhaps even more, than the instrumental features of those decisions, such as their favourability to the individual concerned (Tyler and Blader 2003). In many different settings, people who encounter an authority (for example, the courts, tribunals or the police) pay close attention to how they are treated by its representatives (for example, judges, officials or police officers). This treatment is experienced and assessed independently from the outcome they receive, and people who experience fair treatment during interactions with authorities are more likely to recognize the legitimacy of those authorities, both contextually – during the interaction itself – and in a broader, less bounded sense. Crucially for crime-control policy and wider processes of social regulation, legitimacy then motivates a greater willingness to accept and comply with its decisions, and a greater propensity to cooperate with the authority in the future (Tyler 2006b).

When thinking about procedural justice, it is important to differentiate between:

(a) perceptions/experience of procedural justice enacted by authority figures in a specific encounter;
(b) future expectations of procedural justice that would be enacted by authority figures if one were to come into contact in the future; and
(c) belief that officials can *generally be trusted* to act in procedurally just ways, for example, positive expectations that they *generally* treat people with respect, *generally* treat everybody fairly in one-to-one encounters, and *generally* allow people to have their concerns listened to.

The third belief, that officials generally act in procedurally fair ways, can be framed as perceived trustworthiness because it represents citizens' positive (or negative) expectations regarding valued behaviours (here to act in procedurally fair ways) under conditions of uncertainty (for variations on the theme, see Barber 1983; Gambetta, 1988; Mayer et al 1995; Hardin 2006; Colquitt et al 2007). People cannot know for sure whether others always act fairly and effectively, and to believe that they do is to overcome uncertainty despite imperfect knowledge (Möllering 2001).

This is not to claim that procedural justice is the only important factor of trustworthiness. Effectiveness, distributive justice (that treatment and outcomes are distributed fairly across aggregate social groups) and the idea of bounded authority (that institutional actors do not stray beyond the limits of their rightful authority) may be important additional dimensions of trustworthiness (Trinkner et al 2018; Posch et al 2021; Jackson et al 2022), and may even predominate in some social contexts in shaping trust (willingness to be vulnerable) and legitimacy (Tankebe 2009; Bradford et al 2014). And there are likely a wide range of other factors that are important to different people depending on time and place. The quality of informal social control processes in one's immediate neighbourhood, for example, has been found to be important for trust in the police (Jackson et al 2013), while evaluations of economic performance and indeed people's substantive economic situation have been linked with their readiness to legitimize and trust legal institutions (Bradford and Jackson 2018). That being said, research across multiple contexts points towards the fact that procedural justice is the most consistent, and often most important, predictor of both trust and legitimacy.

The psychological significance of procedural justice

Relational qualities

A reasonable question at this juncture is why procedural justice is such an important antecedent of trust (willingness to be vulnerable) and legitimacy, and indeed why we might expect this to be the case in the context of tribunals and ombuds services. One answer to this question lies in the fundamental importance of fairness – many individuals, in whatever context they are operating, value being treated fairly and respectfully by power-holders they encounter. In the words of Tyler and Nobo (2023: 16): 'people have normative models concerning the ways in which they believe authority should be exercised'. From a normative *and* empirical perspective, fair and respectful process can simply be a 'prescriptive ought' (Folger 2001) – an end in itself and something conceived as essentially important to human relations.

But procedural justice theory also stresses the idea that the importance people place on being treated fairly, stems partly from relational concerns. Of central concern within procedural justice theory are the ways in which cooperation and compliance are motivated and sustained within social groups.

Authorities such as the police and courts (and teachers, employers and indeed parents) represent social groups that many people find important, and the experience of procedural fairness at the hands of representatives of these groups, such as police officers and court officials, communicates messages of (a) inclusion and status within the group and (b) that membership of the group itself is worthwhile (that it is worth belonging to because it treats its members appropriately). When their sense of identification with a group is activated and salient – when they feel included – people are motivated to act in ways that support the group (Tyler and Blader 2003) and to trust and legitimize its authorities. Procedural injustice does precisely the opposite: by communicating a lack of shared values, that the group does not value or care about its (putative) members, it sends the signal that authorities are untrustworthy and illegitimate.

Uncertainty management

Procedural justice can also operate as an uncertainty-reduction mechanism. Procedural justice offers reassurance that an unfamiliar or subjectively unusual procedure is being conducted appropriately and is likely to reach appropriate outcomes (van den Bos and Lind 2002). Indeed, reducing the sense of subjective uncertainty may be *especially important* in situations that people find intimidating, fraught, unsettling and impenetrable (such as formal court or civil proceedings).

Take, for example, the decisions reached in a court. People often do not have the requisite information to properly determine whether an outcome is fair; they are often unsure about how something *actually works*. They want to know whether fair processes are being followed and whether fair outcomes will emerge, so they look at the interpersonal cues of fair treatment and decision-making as indicators one way or the other. In such a context, procedural justice indicates whether the authority figure can be trusted, and thus that the process can be trusted and uncertainty managed. This, coupled with the idea that different people may feel different levels of certainty or uncertainty, suggests that the more uncertain somebody feels, *the more important procedural justice is to them*, and the more important procedural justice is to outcome fairness perceptions.

There are good reasons to believe that relational and uncertainty issues partly overlap in the context of procedural justice. In addition to feeling unsure about how proceedings and processes work, people may also feel unsure about social status and position and therefore look to relational cues that lie at the heart of procedural justice. Structurally, people are ceding authority to another person and system, and this raises the possibility of exploitation and exclusion. So they will be especially sensitive to cues, from procedural justice, that the authority is trustworthy and that their status and value are being affirmed.

Emotions

Emotions are important in people's encounters with criminal and civil justice that are marked by strong asymmetric dependence and power. Power has objective *and* subjective features. On the one hand, power is a structural feature of social relations; it embodies relative differences in the capacity of power-holders to control outcomes that they or others (particularly subordinates) want (Fiske 1993). When encountering legal officials, citizens are dependent on how power and authority are exercised. This can make it an emotionally charged experience, especially for the citizen.

On the other hand, power is a psychological feature of the experience and perception of power of both power-holders and subordinates (Galinsky et al 2006; Smith and Trope 2006). How individuals feel, act and respond partly shapes how the encounter is experienced, how its dynamic develops and the impact it has. If the individual experiences fair interpersonal treatment and fair decision-making processes, this reduces the sense of power distance, helps to generate the sense that power is being rightfully exercised and generates compliance, cooperation and acceptance.

In situations in which people are at the mercy of the power of others, emotions can be positive, especially if one's status is affirmed and uncertainty is being reduced through power-holders acting in procedurally fair ways. But an encounter experienced as procedurally unjust can produce anger, resentment and defiance.

Take, for example, a regulatory encounter like stop-and-search. This is an example of police as 'the sharp edge of governance, the point at which law – as an abstract system or rules – becomes the concrete physical experience of being ruled' (Bowling et al 2019: xii). As Bowling et al (2019: 136): argue: 'Although sometimes thought to be "minimally intrusive", in practice it is often experienced as a highly coercive infringement of liberty.'

Power and authority are palpable in such an encounter. Stop-and-search is a classically low visibility, high discretion, police activity that provides crucial moments in which the legitimacy of the police is asserted, tested and all too often undermined (Geller et al 2014; Tyler et al 2014; Slocum et al 2016). As a key part of the police 'voice' in the legitimacy 'dialogue' envisioned by Bottoms and Tankebe (2012), every stop-and-search encounter involves a claim that police are empowered to treat citizens in this way; that the nature and extent of this power is defensible; and that the ability of police to wield coercive force to ensure compliance is itself justified.

How the police officer acts, and how citizens experience their actions, will be crucial. From a psychological perspective, asymmetric dependence in power situations produces asymmetric social distance between police officers and citizens. For example, because of asymmetric social distance, officers may tend to experience reduced attunement and attentions to citizens they encounter on the street, an increased tendency to objectify, dehumanize and stereotype, and increased expression of interpersonal dominance and aggression (Stamkou et al 2016). Anger may also be used by power-holders to influence the behaviour of subordinates (Clark et al 1996) and, equally, pride and contempt may be used to signal social hierarchy and dissimilarity (Oveis et al 2010).

Procedurally unjust encounters with the police, such as stop-and-search, may create anger and defiance among citizens (Terrill and Mastrofski 2002) as a result of how officers treat them. This could then decrease cooperation and compliance in a way that goes above and beyond the damage that procedural injustice does to trust and legitimacy. Barkworth and Murphy (2015) found that procedurally unjust experiences with police were related to negative affect (for example, anger, anxiety and frustration), which in turn predicted lower levels of compliance (see also Murphy and Tyler 2008). Sargeant et al (2023) linked procedurally unjust encounters with increased identity threat (a sense that the officer was calling into question the individual's dignity and competence) and defiance and resistance towards officers among ethnic minority groups in Australia.

Simply by acting in aggressive, rude and demeaning ways, for example, police officers can be met with resistance and the withdrawal of compliance (Mastrofski et al 2002; McCluskey 2003). This is important because, as Tyler and Nobo (2023: 11) argue:

> If the goal of the police is to ensure compliance, officers must ask how it can be guaranteed. One approach would be to appeal to their legitimate authority. However, if the police lack legitimacy in the eyes of the people they encounter, officers can always default to the use of force to direct public behavior. Most likely, they make this assumption prior to engaging with people, since studies suggest that they typically project dominance immediately in encounters rather than as a reaction to the actions of the civilian (Voigt et al 2017). As noted, this approach further undermines public trust and enhances the need to utilize force once again in the future.

Consider a different sort of police–citizen encounter. Research has shown that emotions and procedural justice can be important, perhaps particularly important, for victims of crime. When a person reports victimization to the police, a range of issues are in play that echo strongly the needs and experiences of those using tribunals and ombuds services. Victims are coping with the mental and sometimes physical effects of the crimes they have experienced; many are vulnerable in multiple ways, and all have turned to 'the authorities' – usually but not always the police – for help and assistance. Their reasons for doing so can range from the transactional (for example, simply requiring a crime reference number for an insurance claim for a stolen bike or particular type of burglary) to the existential, as in cases where immediate protection from an attacker is required.

What victims want from the police thus varies widely. Yet, many accounts stress that victims have strong relational and emotional needs as well as instrumental and practical ones. Alongside physical protection and/or an appropriate bureaucratic response, many victims of crime who report to police seek a sense of reassurance and/or symbolic or psychological redress – that authority sees them as worthy of protection, that their status as rights-bearing citizens has not been fundamentally undermined, and that the loss they have experienced will be restored in ways that do not relate merely to a successful prosecution (or insurance claim).

It is for precisely these reasons that procedural justice is important. Like non-victims, many victims seem to see police as representing 'community' and its values (Elliott et al 2014). For those who experience victimization as alienating and excluding, something that undermines their sense of who they are, police activity can signal (re)integration into, or further exclusion from, society. Two of Elliot et al's (2014: 594) respondents sum up the two sides of this coin:

'They (police) didn't care, they had more important issues like highway patrol; I felt I am not important, not needed in the community.' (#108, victim in a property damage case)

'My wellbeing was important to them (police). That made me feel valued as a member of society.' (#53, assault victim)

Police behaviour that communicates status and inclusion and that the police, and by extension the community, care about the victim – that is, procedural justice – can thus be important as victims start the processes of recovery and of regaining something of what they have lost. It is important to remember that procedural justice carries other information, too, which might also be particularly important for victims – that police are making decisions in an unbiased fashion, are listening to their side of the story, are trustworthy and, perhaps above all, are dealing appropriately with their case (Laxminarayan 2012; Koster et al 2020). This helps the victim manage the process uncertainty they almost inevitably feel and generates a feeling they have some level of control over what is happening to them; both of which experiences, in line with the classic formulations of procedural justice theory, can offer reassurance that the right outcome has been, or will be, reached.

Several studies have stressed the negative effects of police treatment experienced as procedurally unjust. This evokes anger, frustration and resentment (Laxminarayan 2012; Wemmers 2013; Barkworth and Murphy 2015), with implications for psychological wellbeing. Barkworth and Murphy (2015) link this explicitly to status uncertainty – victims who feel unfairly treated by police feel less secure in themselves and in society, triggering a negative emotional response in defence that, in turn, diminishes their quality of life.

Moreover, a significant body of research in organizational settings has linked procedural (in)justice to physical health outcomes. Here, procedural injustice is linked to feelings of powerlessness, confusion and anxiety (Eib et al 2018), which in turn elicit a sense of threat, a 'flight or fight' response, and thus stress or 'allostatic overload' (McEwen 2008). Through this and related mechanisms, experiences of procedural injustice have been linked to health outcomes such as depression and psychiatric morbidity, sickness-related work absence and even coronary heart disease (Kivimäki et al 2005, 2007; Ferrie et al 2006; Grynderup et al 2013; Eib et al 2018). While, as noted previously, this research has concentrated on organizational settings, and thus pertains to the employee/employer relationship, it is at least plausible to suggest that something similar might be found in the victim/police relationship. Notably, many victims are already highly stressed before they contact police, and poor police practice may thus exacerbate already existing factors. And, again, what pertains in the victim/police relationship might also pertain in the relationship between services users and tribunals or ombuds procedures.

So what about the administrative justice system?

In interactions with tribunals and ombuds, emotions may be important, especially in the context of frustration and anxiety in the wake of a lack of clarity about *how things work*. For help-seekers, emotions may be less about anger and more about dismay, upset and disappointment, and more about it being emotionally rewarding to have status recognition and to 'say one's peace'. Yet, due to the dearth of UK-based research on how people use and think about tribunals and ombuds services, open empirical questions remain, particularly in the new era of mediated interaction. While Creutzfeldt and Bradford (2016) identified a 'procedural justice effect' in their UK data, drawn from users of ombuds services, their findings related to a combination of online, traditional telephone, written and, to an extent, face-to-face interactions. As a result, it is unclear how this translates into 'only' online interactions. Procedural justice may be less important in the context of the financial and usually very outcome-focused nature of these interactions, where most service users are clear what they want, and what 'success' looks like to them (indeed, having easily identifiable and quantifiable outcomes such as renegotiating a rent increase may mean that process concerns are suppressed, precisely because the outcome is so easy to grasp).

The relational aspects of procedural justice seem to hold firm in many different group contexts – where, for example, the group is defined in ways ranging from the nation to the family (Fondacaro et al 1998). There is no particular reason to suggest this will not be the case in the present context. Judges and ombuds can be construed as authority figures of particular groups, most obviously the legal 'community' – or state – that mandates their activities and to the proper function of which their activities contribute. But there are other reasons why procedural justice concerns are so central to people's experiences of legal processes, and these might also be important in civil courts and ADR processes.

Our qualitative exploration of emotions

In the qualitative work presented in Chapters 7 and 8, we explore procedural justice, drawing out emotions at various junctures. In the earliest iterations of procedural justice theory (Thibaut and Walker 1975; Walker et al 1979), process fairness was construed as important because it gave those involved in court processes a sense that they had some measure of input into and therefore control over the procedure concerned. Having 'voice' in a procedure offers reassurance that the right outcome has been, or is likely to be, reached. In this early work, the idea of procedural justice had a more instrumental flavour than in the relational account outlined in the previous section – and it would seem clear that those using an ADR option have an interest in the outcome reached. Moreover, 'outcome' can refer to the specific settlement reached *and/or* its fairness in distributive terms, that is, in terms of whether others, in a similar situation, received a similar outcome. It follows that those involved in legal processes may infer both the quality of an outcome and its fairness in distributive terms from the procedural justice they experience. In situations where the decision-making process is hard to understand, opaque or even hidden from participants, they may be especially attuned to process fairness because they have few other ways of coming to a judgement about the overall quality of the decision-making or the fairness of outcomes reached.

Building on this, it can be suggested that procedural justice, and the trust and legitimacy it engenders, may be particularly important during times of turbulence and change, such as for instance ADR shifting quickly online. One of the reasons for the importance people place on procedural justice in their interactions with authorities is that it acts as an uncertainty-reduction mechanism, offering reassurance that an unfamiliar or subjectively unusual procedure is being conducted appropriately (and is likely to reach appropriate outcomes) (van den Bos and Lind 2002). Moreover, when people are uncertain, unclear or simply unfamiliar with a process or situation, they look for information that their status and right to be in that situation, at least, is affirmed by the authorities they are dealing with; procedural justice is also important for the relational signals of belonging and inclusion it sends to people (Tyler and Blader 2000, 2003). All of this, coupled with the weight of research identifying procedural justice effects across multiple justice settings, implies that we might expect procedural justice to be central to the judgements people make about online tribunals and ombuds hearings. These situations are, after all, novel in their use of digitally mediated interactions, and relatively unusual from the perspective of most users, who are unfamiliar with civil justice and ADR.

Our quantitative exploration

In the quantitative work presented in Chapter 6, we explore procedural justice, legitimacy and outcome fairness. We focus, among other things, on the relationship between procedural justice and perceptions of outcome fairness. The procedural justice literature has long been interested in people's perceptions of outcome fairness and favourability. In their foundational work, Thibault and Walker (1975; Walker and Thibault 1978) viewed both outcome control (control over the actual decision made) and process control (essentially a form of voice, for example, that people have a say and can present their own side of the story) to be at the heart of procedural justice. They argued that the procedures used to make decisions in courts were important to people's overall fairness perceptions and independent of the favourability of the outcome. Moreover, procedural justice was important precisely because it allows people to feel like they can shape, or have a say in, the final outcome.

The relational/group–value/group engagement model (Lind and Tyler 1988; Tyler and Lind 1992) takes a different view. People care about procedural justice not so much because they have a 'dog in the race' (they want to be able to get a 'good' outcome), but rather because procedures give important information about how they are viewed by group authorities, shaping their sense of self-worth and standing within the group. Moreover, Tyler (1990) refers to procedural justice as a 'reservoir of support' rooted in deference and authorization: not only do people care more about process than outcome (because relational signals are more important than instrumental outcomes), but when people are certain that they have been treated fairly (for example, the decision-making process was unbiased, transparent and accountable) they are more likely to believe that the outcome was fair and independent whether it was favourable or not.

Fairness heuristic theory (van den Bos et al 1997, 1998) and uncertainty management theory (van den Bos 2001; Lind and van den Bos 2002; van den Bos and Lind 2002) add a complementary picture. The basic idea is that people often do have the requisite information to properly determine whether an outcome is fair. In such a context, procedural justice acts as an uncertainty-reduction mechanism, precisely because it indicates whether the authority figure can be trusted. Procedural justice offers reassurance that an unfamiliar or subjectively unusual procedure is being conducted appropriately (and is likely to reach appropriate outcomes) (van den Bos and Lind 2002).

Moreover, people may feel unsure about their social status and position and therefore look to relational cues that lie at the heart of procedural justice. Structurally, people are ceding authority to another person, and this raises the possibility of exploitation and exclusion. So people are especially attuned to cues from procedural justice that the authority is trustworthy. They may also feel unsure about how something *actually works*. They may want to know whether fair processes are being following and whether fair outcomes will emerge, and they may look at the interpersonal cues of fair treatment and decision-making as indicators one way or the other. This, coupled with the idea that different people may feel different levels of certainty or uncertainty, suggests that the more uncertain somebody feels, *the more important procedural justice is to them*, and the more important procedural justice is to transparency and outcome fairness perceptions.

Conclusion

This chapter has considered the crucial role that procedural justice plays in fostering public trust and institutional legitimacy. Drawing upon a rich tradition in the fields of policing and criminal justice, we have introduced these concepts and highlighted their psychological significance. Moreover, we have examined how these principles can be applied to the administrative justice system (AJS), particularly in the context of relational qualities, uncertainty management and emotions. By considering these aspects, we have established a framework that, in conjunction with the insights presented in Chapter 1, provides a comprehensive lens to understand the empirical chapters in Part III of the book. This framework allows us to analyse and interpret the empirical findings within the broader context of procedural justice, public trust and institutional legitimacy. Through this integrated approach, we aim to deepen our understanding of the intricate dynamics at play and shed light on the implications for the AJS.

The next part of the book introduces pathways to justice in three chapters. Chapter 3 outlines the two areas of law we are concerned with, housing and special educational needs and disabilities (SEND). Chapter 4 is about housing and Chapter 5 about SEND, following the help-seekers' ideal case pathways through institutions of the AJS, as well as revealing what our data showed about these pathways.

References

Barber, B. (1983) *The Logic and Limits of Trust*, New Brunswick, NJ: Rutgers University Press.

Barkworth, J.M. and Murphy, K. (2015) 'Procedural justice policing and citizen compliance behaviour: the importance of emotion', *Psychology, Crime & Law* 21(3): 254–273.

Bottoms, A. and Tankebe, J. (2012) 'Beyond procedural justice: a dialogic approach to legitimacy in criminal justice', *Journal of Criminological Law and Criminology* 102: 119–170.

Bolger, P.C. and Walters, G.D. (2019) 'The relationship between police procedural justice, police legitimacy, and people's willingness to cooperate with law enforcement: a meta-analysis', *Journal of Criminal Justice* 60: 93–99. https://doi.org/10.1016/j.jcrimjus.2019.01.001

Bowling, B., Reiner, R. and Sheptycki, J.W. (2019) *The Politics of the Police*, Oxford: Oxford University Press.

Bradford, B. and Jackson, J. (2018) 'Police legitimacy among immigrants in Europe: institutional frames and group position', *European Journal of Criminology* 15: 567–588.

Bradford, B., Murphy, K. and Jackson, J. (2014) 'Officers as mirrors: policing, procedural justice and the (re)production of social identity', *British Journal of Criminology* 54(4): 527–550.

Bradford, B., Creutzfeldt, N. and Steffek, F. (2023) 'Thinking holistically about procedural justice in alternative dispute resolution: a case study of the German Federal Ombudsman Scheme', *Law & Social Inquiry* 48(3): 748–779. doi:10.1017/lsi.2022.5

Clark, M.S., Pataki, S.P. and Carver, V.H. (1996) 'Some thoughts and findings on self-presentation of emotions in relationships', in G.J.O. Fletcher and J. Fitness (eds), *Knowledge Structures in Close Relationships: A Social Psychological Approach*, Mahwah, NJ: Erlbaum, 247–274.

Colquitt, J.A., Scott, B.A. and LePine, J.A. (2007) 'Trust, trustworthiness, and trust propensity: a meta-analytic test of their unique relationships with risk taking and job performance', *Journal of Applied Psychology* 92: 909–927.

Creutzfeldt, N. and Bradford, B. (2016) 'Dispute resolution outside of courts: procedural justice and decision acceptance among users of ombuds services in the UK', *Law & Society Review* 50(4): 985–1016.

Eib, C., Bernhard-Oettel, C., Magnusson Hanson, L.L. and Leineweber, C. (2018) 'Organizational justice and health: studying mental preoccupation with work and social support as mediators for lagged and reversed relationships', *Journal of Occupational Health Psychology* 23(4): 553–567.

Elliott, I., Thomas, S. and Ogloff, J. (2014) 'Procedural justice in victim-police interactions and victims' recovery from victimisation experiences', *Policing and Society* 24(5): 588–601.

Ferrie, J., Head, J., Shipley, M., Vahtera, J., Marmot, M. and Kivimäki, M. (2006) 'Injustice at work and incidence of psychiatric morbidity: the Whitehall II study', *Occupational and Environmental Medicine* 63(7): 443–450.

Fiske, S.T. (1993) 'Controlling other people: the impact of power on stereotyping', *American Psychologist* 48: 621–628.

Folger, R. (2001) 'Fairness as deonance', in S.W. Gilliland, D.D. Steiner and D.P. Skarlicki (eds), *Theoretical and Cultural Perspectives on Organizational Justice*, Greenwich, CT: Information Age Publishing, 3–33.

Fondacaro, M.R., Dunkle, M.E. and Pathak, M.K. (1998) 'Procedural justice in resolving family disputes: a psychosocial analysis of individual and family functioning in late adolescence', *Journal of Youth and Adolescence* 27: 101–119.

Galinsky, A.D., Magee, J.C., Inesi, M.E. and Gruenfeld, D.H. (2006) 'Power and perspectives not taken', *Psychological Science* 17: 1068–1074.

Gambetta, D. (1988) *Trust: Making and Breaking Cooperative Relations*, Oxford: Basil Blackwell.

Geller, A., Fagan, J., Tyler, T. and Link, B.G. (2014) 'Aggressive policing and the mental health of young urban men', *American Journal of Public Health* 104(12): 2321–2327.

Grynderup, M.B., Mors, O., Hansen, Å.M., Andersen, J.H., Bonde, J.P., Kærgaard, A. and Kolstad, H.A. (2013) 'Work-unit measures of organisational justice and risk of depression – a 2-year cohort study', *Occupational and Environmental Medicine* 70(6): 380–385.

Hamm, J.A., Trinkner, R. and Carr, J.D. (2017) 'Fair process, trust, and cooperation: moving toward an integrated framework of police legitimacy', *Criminal Justice and Behavior* 44(9): 1183–1212.

Hamm, J.A., Wolfe, S.E., Cavanagh, C. and Lee, S. (2022) '(Re) organizing legitimacy theory', *Legal and Criminological Psychology* 27(2): 129–146.

Hardin, R. (2002) *Trust and Trustworthiness*, New York: Russell Sage Foundation.

Hardin, R. (2006) *Trust and Trustworthiness*, New York: Russell Sage Foundation.

Hough, M. (2020) *Good Policing: Trust, Legitimacy and Authority* (online edn), Bristol: Policy Press Scholarship Online. https://doi.org/10.1332/policypr ess/9781447355076.001.0001

Jackson, J. and Gau, J.M. (2016) 'Carving up concepts? Differentiating between trust and legitimacy in public attitudes towards legal authority', in In E. Shockley, T.M.S. Neal, L.M. PytlikZillig and B.H. Bornstein (eds), *Interdisciplinary Perspectives on Trust: Towards Theoretical and Methodological Integration*, Cham: Springer, 49–69.

Jackson, J., Bradford, B., Hough, M., Myhill, A., Quinton, P. and Tyler, T.R. (2012) 'Why do people comply with the law? Legitimacy and the influence of legal institutions', *British Journal of Criminology* 52(6): 1051–1071.

Jackson J., Bradford B., Stanko B. and Hohl, K. (2013) *Just Authority? Trust in the Police in England and Wales*, New York: Routledge.

Jackson, J., Pósch, K., Oliveira, T.R., Bradford, B., Mendes, S.M., Natal, A.L. and Zanetic, A. (2022) 'Fear and legitimacy in São Paulo, Brazil: police–citizen relations in a high violence, high fear city', *Law & Society Review* 56(1): 122–145.

Kivimäki, M., Ferrie, J.E., Brunner, E., Head, J., Shipley, M.J., Vahtera, J. and Marmot, M.G. (2005) 'Justice at work and reduced risk of coronary heart disease among employees: the Whitehall II Study', *Archives of Internal Medicine* 165(19): 2245–2251.

Kivimäki, M., Vahtera, J., Elovainio, M., Virtanen, M. and Siegrist, J. (2007) 'Effort-reward imbalance, procedural injustice and relational injustice as psychosocial predictors of health: complementary or redundant models?', *Occupational and Environmental Medicine* 64(10): 659–665.

Koster, N.S.N., van der Leun, J.P. and Kunst, M.J. (2020) 'Crime victims' evaluations of procedural justice and police performance in relation to cooperation: a qualitative study in the Netherlands', *Policing and Society* 30(3): 225–240.

Laxminarayan, M. (2012) 'Procedural justice and psychological effects of criminal proceedings: the moderating effect of offense type', *Social Justice Research* 25: 390–405.

Lind, E.A. and Tyler, T.R. (1988) *The Social Psychology of Procedural Justice*, New York: Springer Science + Business Media.

Lind, E.A. and van den Bos, K. (2002) 'When fairness works: toward a general theory of uncertainty management', *Research in Organizational Behavior* 24: 181–223.

Mastrofski, S.D., Reisig, M.D. and McCluskey, J.D. (2002) 'Police disrespect toward the public: an encounter-based analysis', *Criminology* 40(3): 519–552.

Mayer, R.C., Davis, J.H. and Schoorman, F.D. (1995) 'An integrative model of organizational trust', *Academy of Management Review* 20: 709–734.

Mazerolle, L., Bennett, S., Davis, J., Sargeant, E. and Manning, M. (2013) 'Procedural justice and police legitimacy: a systematic review of the research evidence', *Journal of Experimental Criminology*, 9(3): 245–274. https://doi.org/10.1007/s11292-013-9175-2

McCluskey, J.D. (2003) *Police Requests for Compliance: Coercive and Procedurally Just Tactics*, El Paso, TX: LFB Scholarly.

McEwen, B.S. (2008) 'Central effects of stress hormones in health and disease: understanding the protective and damaging effects of stress and stress mediators', *European Journal of Pharmacology* 583(2–3): 174–185.

Möllering, G. (2001) 'The nature of trust: from Georg Simmel to a theory of expectation, interpretation and suspension', *Sociology* 35(2): 403–420.

Murphy, K. and Tyler, T. (2008) 'Procedural justice and compliance behaviour: the mediating role of emotions', *European Journal of Social Psychology* 38: 652–668.

Nagin, D.S. and Telep, C.W. (2017) 'Procedural justice and legal compliance', *Annual Review of Law and Social Science* 13: 5–28.

Nagin, D.S. and Telep, C.W. (2020) 'Procedural justice and legal compliance: a revisionist perspective', *Criminology and Public Policy* 19(3): 761–786. https://doi.org/10.1111/1745-9133.12499

Oveis, C., Horberg, E.J. and Keltner, D. (2010) 'Compassion, pride, and social intuitions of self-other similarity', *Journal of Personality and Social Psychology* 98: 618–630.

Pass, M.D., Madon, N.S., Murphy, K. and Sargeant, E. (2020) 'To trust or distrust? Unpacking ethnic minority immigrants' trust in police', *The British Journal of Criminology* 60(5): 1320–1341.

Pósch, K., Jackson, J., Bradford, B. and Macqueen, S. (2021) '"Truly free consent"? Clarifying the nature of police legitimacy using causal mediation analysis', *Journal of Experimental Criminology* 17: 563–595.

PytlikZillig, L.M. and Kimbrough, C.D. (2016) 'Consensus on conceptualizations and definitions of trust: Are we there yet?', in E. Shockley, T.M.S. Neal, L.M. PytlikZillig and B.H. Bornstein (eds), *Interdisciplinary Perspectives on Trust: Towards Theoretical and Methodological Integration*, Cham: Springer, 17–47.

Sargeant, E., Murphy, K. and Bradford, B. (2023) 'The foundations of defiance: examining the psychological underpinnings of ethnic minority defiance toward police', *Policing and Society* 33: 802–819.

Slocum, L.A., Ann Wiley, S. and Esbensen, F.A. (2016) 'The importance of being satisfied: a longitudinal exploration of police contact, procedural injustice, and subsequent delinquency', *Criminal Justice and Behavior* 43(1): 7–26.

Smith, P.K. and Trope, Y. (2006) 'You focus on the forest when you're in charge of the trees: power priming and abstract information processing', *Journal of Personality and Social Psychology* 90: 578–596.

Stamkou, E., van Kleef, G.A., Fischer, A.H. and Kret, M.E. (2016) 'Are the powerful really blind to the feelings of others? How hierarchical concerns shape attention to emotion', *Personal Social Psychology Bulletin* 42: 755–768.

Sunshine, J. and Tyler, T.R. (2003) 'The role of procedural justice and legitimacy in shaping public support for policing', *Law & Society Review* 37(3): 513–548.

Tankebe, J. (2009) 'Public cooperation with the police in Ghana: does procedural fairness matter?' *Criminology* 47(4): 1265–1293.

Terrill, W. and Mastrofski, S.D. (2002) 'Situational and officer-based determinants of police coercion', *Justice Quarterly* 19(2): 215–248.

Thibaut, J. and Walker, L. (1975) *Procedural Justice: A Psychological Analysis*, Hillsdale, NJ: Erlbaum.

Trinkner, R., Jackson, J. and Tyler, T.R. (2018) 'Bounded authority: expanding "appropriate" police behavior beyond procedural justice', *Law and Human Behavior* 42: 280–293.

Tyler, T.R. (1990) 'Justice, self-interest, and the legitimacy of legal and political authority', in J.J. Mansbridge (ed), *Beyond Self-interest*, Chicago: University of Chicago Press, 171–179.

Tyler, T.R. (2003) 'Procedural justice, legitimacy, and the effective rule of law', *Crime and Justice* 30: 283–357. http://www.jstor.org/stable/1147701

Tyler, T.R. (2006a) 'Psychological perspectives on legitimacy and legitimation', *Annual Review of Psychology* 57(1): 375–400. https://doi.org/10.1146/annurev.psych.57.102904.190038

Tyler, T.R. (2006b) *Why People Obey the Law* (2nd edn), New Haven, CT: Yale University Press.

Tyler, T.R. (2017) 'Procedural justice and policing: a rush to judgment?', *Annual Review of Law and Social Science* 13: 29–53. https://doi.org/10.1146/annurev-lawsocsci-110316-113318

Tyler, T.R. and Blader, S.L. (2000) *Cooperation in Groups: Procedural Justice, Social Identity, and Behavioral Engagement*, London: Psychology Press.

Tyler, T.R. and Blader, S.L. (2003) 'The group engagement model: procedural justice, social identity, and cooperative behavior', *Personality and Social Psychology Review* 7: 349–561.

Tyler, T.R. and Fagan, J. (2008) 'Legitimacy and cooperation: why do people help the police fight crime in their communities?', *Ohio State Journal of Criminal Law* 6: 231–275.

Tyler, T.R. and Huo, Y.J. (2002) *Trust in the Law: Encouraging Public Cooperation with the Police and Courts*, New York: Russell Sage Foundation.

Tyler, T.R. and Jackson, J. (2014) 'Popular legitimacy and the exercise of legal authority: motivating compliance, cooperation, and engagement', *Psychology, Public Policy, and Law* 20(1): 78–95.

Tyler, T.R. and Lind, E.A. (1992) 'A relational model of authority in groups', *Advances in Experimental Social Psychology* 25: 115–191.

Tyler, T.R. and Nobo, C. (2023) *Legitimacy-Based Policing and the Promotion of Community Vitality*, Cambridge: Cambridge University Press.

Tyler, T.R., Fagan, J. and Geller, A. (2014) 'Street stops and police legitimacy: teachable moments in young urban men's legal socialization', *Journal of Empirical Legal Studies* 11: 751–785.

van den Bos, K. (2001) 'Uncertainty management: the influence of uncertainty salience on reactions to perceived procedural fairness', *Journal of Personality and Social Psychology* 80(6): 931–941.

van den Bos, K. and Lind, E.A. (2002) 'Uncertainty management by means of fairness judgments', *Advances in Experimental Social Psychology* 34: 1–60. https://doi.org/10.1016/S0065-2601(02)80003-X

van den Bos, K., Vermunt, R. and Wilke, H.A.M. (1997) 'Procedural and distributive justice: what is fair depends more on what comes first than on what comes next', *Journal of Personality and Social Psychology* 72(1): 95–104.

van den Bos, K., Wilke, H.A. and Lind, E.A. (1998) 'When do we need procedural fairness? The role of trust in authority', *Journal of Personality and Social Psychology* 75(6): 1449–1458.

Voigt, R., Camp, N.P., Prabhakaran, V., Hamilton, W.L., Hetey, R.C., Griffiths, C.M., et al (2017) 'Language from police body camera footage shows racial disparities in officer respect', Proceedings of the National Academy of Science USA 114(25): 6521–6526. doi: 10.1073/pnas.1702413114.

Walker, L. and Thibault, J. (1978) 'A theory of procedure', *California Law Review* 66: 544–545.

Walker, L., Lind, E.A. and Thibaut, J. (1979) 'Relation between procedural and distributive justice', *Virginia Law Review* 65: 1401.

Walters, G.D. and Bolger, P.C. (2019) 'Procedural justice perceptions, legitimacy beliefs, and compliance with the law: a meta-analysis', *Journal of Experimental Criminology* 15: 341–372.

Wemmers, J.A. (2013) 'Victims' experiences in the criminal justice system and their recovery from crime', *International Review of Victimology* 19(3): 221–233.

PART II

Pathways to Justice

3

Two Areas of Law in Context and the Help-Seeker Journey

Introduction

This book is about understanding how people accessed the administrative justice system (AJS) during the pandemic and what lessons we can learn as we move on from the coronavirus crisis. Following the two theoretical chapters in Part I, this part offers an introduction to the two pathways to justice in this book. This chapter provides an overview of the help-seeker journey. The help-seeker journey follows the person with a problem and legal need through different stages of seeking help, finding advice and reaching an ombuds or tribunal to resolve their problem. For housing, we look at the advice sector, the Property Chamber and the Housing Ombudsman; and for SEND we look at the advice sector, the SEND Tribunal, the Local Government and Social Care Ombudsman (LGSCO) and the Parliamentary and Health Service Ombudsman (PHSO). The emphasis on these institutions allows us to understand in some depth the effects of the pandemic, how they managed to provide their services remotely, and what lessons can be learned for the AJS and the justice system more generally.

Existing research based on legal needs has demonstrated that those experiencing the greatest social and economic disadvantage and marginalization are often the least likely to take any action in response to a rights-based problem (Gibson and Caldeira 1996; Hazel and Beinart 1999; Hertogh 2009; Gill and Creutzfeldt 2018), particularly those people who do nothing in response to a problem experienced, which is relatively common in both housing and SEND contexts.

The impact of COVID-19 on the AJS and its users has been significant. Digitalization by default, remote hearings and buildings not being physically accessible have brought many challenges. Access to advice, to support and to justice has been compromised, especially for marginalized groups (those without, for example, reliable or indeed any internet connection).

The nature of the lockdown responses to the pandemic meant that digital systems had to operate in a far from normal environment. Recent reports[1] (LEF 2020; Tomlinson et al 2020) show how parts of the justice system are coping under the pandemic.

In the following sections, we provide the context and policy background for housing and SEND (in preparation for Chapters 4 and 5) and then we map the pathways to resolve housing and SEND disputes through the lens of the help-seeker journey as an *ideal case scenario*. We build on this in Chapters 4 and 5 with the help of our empirical data to show how these ideal case scenarios pan out in *actual cases* drawn from our dataset.

Housing

The wider context: the housing crisis and the impact of the pandemic

Housing in the UK – particularly in London and the south-east of England – is some of the most expensive and cramped in the world.[2] Housing costs in the UK are not only high in absolute terms but also relative to incomes, and UK house prices are not only extraordinarily high but also exceptionally volatile. The current housing affordability crisis has been developing slowly over the last 40 years (Hilber and Schoni 2016).

A report published in 2018 by the housing and homelessness charity Shelter showed in vivid detail the housing crisis as it exists in England. It is a crisis principally of those who rent, not through choice, but because of the unaffordability of housing for would-be homeowners. This has left millions in insecure and expensive rented accommodation. The report found that most private renters on low incomes struggle to afford their rent, with private renters entitled to little legal protection from eviction. In addition, private renters often face threats to health and safety, having to tackle problems with their homes that include electrical hazards, damp and pest infestation (Creutzfeldt et al 2021). If private renters make a formal complaint about their housing issue, research suggests there's a 50:50 chance they will be handed an eviction notice within six months. Moreover, the report explained how stigma and discrimination are linked to housing in a number of ways. Stigma can make social renters feel powerless to influence decisions about their homes; and in the private market, refusing to rent homes to those receiving benefits is widespread. A further report published by Shelter in December 2021 demonstrated that, in the context of this housing emergency, in the UK 17.5 million people are denied the right to a safe home and women are disproportionately affected (Schofield 2021).

The roots of the housing crisis track back to the decline of social housing over the last 40 years, coupled with the trends of rising prices, falling ownership and an expanding – but increasingly unfit – private rented sector, paid for by a rapidly rising housing benefit Bill (Shelter 2018).

At the sharpest end of the crisis, more and more people are being left homeless. An alarming 282,000 single people, couples and families were homeless or threatened with homelessness by local authorities (LAs) in 2020/ 2021 in England (Crisis 2022).

The COVID-19 pandemic has severely disrupted construction, made it difficult for many households to pay for shelter, and seriously hurt the housing sector (Organisation for Economic Co-operation and Development 2020).[3] Crisis (2022) conducted a longitudinal study of the impacts of recent economic and policy developments on homelessness in England. Several key findings are relevant here. Progress against the government's target of ending rough sleeping by 2024, supported by substantially increased investment, including via the Rough Sleeping Initiative, has been radically accelerated by responses to the pandemic. Total temporary accommodation placements continued to increase (up by 4% in 2020/2021), and bed-and-breakfast hotel placements rose significantly (by 37%). Some of this increase reflects actions under the *Everyone In* programme (with clear reductions in rough sleeping (down 33%) and sofa surfing (down 11%)), though temporary accommodation placements were already on an upward trajectory before the pandemic. Related to this, most LA survey respondents (78%) also reported that access to private rented sector accommodation became more difficult during 2020/2021, with 57% identifying access to the social rented sector as becoming more challenging also.

COVID-19 inflicted considerable damage on the economy during 2020, and although 2021 saw some bounce back, uncertainty remains regarding when and how the economy will recover following the pandemic. Government plans to increase spending on public services, including health and local government, will depend on the performance of the economy and pandemic-related developments. Uncertain economic prospects and the deepening living-cost crisis led to mounting concerns that there may be a surge in homelessness in 2022/2023. It is predicted that the aftermath of the COVID-19 pandemic risks a substantial rise in core homelessness, with overall levels expected to sit one-third higher than 2019 levels on current trends. Levels of rough sleeping are also predicted to rise, despite the government's target of ending this form of homelessness by 2024 (Crisis 2022).

In light of the aforementioned, experts in the field argue that during this period neither the government nor any opposition party has put forward any effective policy proposals to tackle the housing crisis in the UK (LSE 2019). This has been argued pre-COVID and there was no effective plan, which now has been compounded and made more urgent by the fallout from COVID.

In the next section we turn to an examination of major housing reforms and Acts.

Policy developments

Housing is one of the government's key priorities. For many people, the availability and affordability of housing has become increasingly difficult in recent years. Major housing reforms and Acts have been introduced to try to support people through the housing crisis faced by those residing in England. It is to these that we now turn.

Government involvement in housing encompasses a diverse array of organizations, individuals and activities: government departments – housing policy is led by the Department for Communities and Local Government but a range of other departments are involved (for example, Department for Work and Pensions); various organizations and individuals, including housing developers, building contractors, mortgage lenders, LAs, housing associations, landlords, owner-occupiers, private renters and those in the social rented sector; and interventions, including regulation (for example, planning), grant funding (for example, for new homes), and loans (for example, help-to-buy equity loans) (National Audit Office 2017).

The key housing policies that were adopted in the past, especially those implemented in more recent years, not surprisingly reflect the fact that housing affordability has been a significant concern of voters and politicians. Hilber and Schoni (2016) discuss in detail the UK's key policies that were implemented up until 2016 with the intent of addressing the affordability crisis, assessing policy objectives, as well as their merits and demerits. At the threshold, this concerns (a) social housing, (b) right-to-buy, (c) help-to-buy and (d) housing-related tax policies. The provision of social housing has certainly helped the lowest-income households and the most vulnerable people to obtain more adequate housing than they could have in the absence of such intervention. However, policies related to social housing, such as the so-called 'section 106 agreements', which require private-sector developers to offer 'affordable housing' as a condition of obtaining planning permission, have adverse effects on social housing because the demand for such subsidized housing far outstrips supply.

The downturn of social housing began in 1980, when Margaret Thatcher introduced the 'right-to-buy'. In brief, the policy allows social tenants to purchase their homes at a significantly subsidized price, with the effect that some of the best social housing stock moved from socially rented to privately owned. Right-to-buy is a crucial factor helping to explain the significant rise in homeownership from 1980 until 2002. The so-called 'help-to-buy' policy was introduced in 2013. The aim of the scheme had been to stimulate housing demand. The help-to-buy scheme consists of four instruments: equity loans, mortgage guarantees, shared ownership and a 'new buy' scheme that allows buyers to purchase a newly built home with a deposit of only 5% of the purchase price. The promoters of the policy hoped that the increase in demand would translate into new housing being supplied and higher homeownership attainment. Finally, the key housing-related taxes in the UK include central government grants to LAs and the council tax, within which most local expenditures in the UK are financed via central government grants, not via local taxes.

The Grenfell Tower fire in June 2017 exposed a range of issues with social housing and provided an impetus for change. In August 2018, following extensive engagement and consultation with social housing residents across the country, the government published a social housing green paper – 'A new deal for social housing' (Ministry of Housing, Communities & Local Government 2018a) – which aimed to 'rebalance the relationship between residents and landlords'. Alongside the green paper, the government published a call for evidence for a review of social housing regulation (Ministry of Housing, Communities & Local Government 2018b) which sought views on how well the regulatory regime was operating. The consultation ran from 14 August to 6 November 2018 and received over 1,000 responses. After a gap of two years, on 17 November 2020 the government published a social housing white paper – the Charter for Social Housing Residents (Ministry of Housing, Communities and Local Government 2020). The Charter sets out measures designed to deliver on the government's commitment to the Grenfell community that 'never again would the voices of residents go unheard' and on its 2019 manifesto pledge,[4] to empower residents, provide greater redress, better regulation and improve the quality of social housing. The white paper is intended to deliver 'transformational change' for social housing residents. It sets out measures to, among other things, ensure social housing is safe and of good quality, ensure swift and effective complaint resolution, and empower and support residents in engaging with and holding their landlords to account – which entails changes to the Housing and Planning Act 2016 that contains the main housing laws and legislations in the UK (House of Commons 2022).

Although the white paper has been well received by tenants, social landlords and the housing sector, concerns have been expressed about the slow pace of social housing reform, the failure to address issues around the supply of homes for social rent, the lack of clarity about whom and what social housing is for, the failure to fully address the issue of stigma (exacerbated by the government's strong focus on home ownership), the lack of a national platform or representative body to represent tenants' interests, and potential challenges for social landlords in resourcing all the new requirements.[5] Furthermore, there is no timetable attached to delivering the measures set out in the social housing white paper (House of Commons 2022).

Landlords, tenants and other court users in housing cases have raised issues and concerns about current court and tribunal processes. Currently, tenants, landlords and property agents can bring a range of housing issues to the courts or the Property Tribunal to enforce their rights. Users have expressed disquiet that processes do not always work as effectively and as efficiently as they could. Tenants and landlords experience difficulties in navigating the process of bringing a case to a court and have found that the length of time the process takes can make it costly and demanding. In October 2017, the government committed to consult with the judiciary on whether a new, specialist housing court could help to address these concerns. Having considered the responses to the 2018 call for evidence and associated stakeholder engagement, the conclusion in 2022 was that the costs of introducing a new housing court would outweigh the benefits, and that there are more effective and efficient ways to address the issues experienced by court and tribunal users in housing cases. Therefore, the government is implementing a package of wide-ranging reforms instead.[6]

In June 2022, the government published its Social Housing Regulation Bill[7] – as a direct result of the Grenfell Tower fire – putting into law a host of reforms to the regulation of the sector. Key changes included the following measures:

- the Regulator of Social Housing (RSH) will be required to set up an advisory panel, made up of representatives of social housing tenants, social landlords and their lenders, councils, the Greater London Authority, Homes England and the housing secretary;
- underperforming social landlords will be subject to inspections;
- the RSH has been granted powers to issue social landlords with 'performance improvement plan notices' if they fail to meet standards, if there is a risk they will fail to meet standards, and if they fail to provide documents or information the RSH has asked for;
- the RSH has the power to carry out emergency works on properties, for which the social landlord will have to pay;
- there will be mandatory checks on electrical installations at least every five years for rented and leasehold properties, and mandatory portable appliance testing on all electrical appliances that are provided by social landlords;
- every registered provider will have to appoint a health and safety lead;
- housing associations will now be subject to a freedom of information-style information-sharing process;
- the 'serious detriment' test has been removed, which blocks the RSH from intervening over consumer standards unless it suspects tenants are at risk of serious harm;
- the regulator can now ask social landlords to collect and publish information relating to their compliance performance; and
- finally, critical to the purposes of this book, the Bill proposes an improved Housing Ombudsman Scheme.

The Housing Ombudsman was granted new powers – which included the ability to refer more cases to the regulator and to issue complaint-handling orders against poorly performing landlords – in September 2020. The purpose of a complaint-handling failure order is to ensure that a landlord's complaint-handling process is accessible and consistent, and that it enables the timely progression of complaints for residents, as set out in the Housing Ombudsman's complaint-handling code. The Bill puts into law the code of practice. It also legally allows the watchdog to order a landlord to review its policies on specific issues. Added to this, the Ombudsman and the RSH must, by law, prepare and maintain a memorandum describing how they intend to work together as they perform their duties.

Other key housing policies have focused on homelessness amid concerns of the rising levels of homeless people in the UK. For example, the Homelessness Reduction Bill was introduced in the House of Commons on 29 June 2016 (Local Government Association 2017). The ambition of the Homeless Reduction Act is to shift the culture of homelessness services towards prevention and provide assistance to all eligible people in need of it, removing barriers for service users (Ministry of Housing, Communities and Local Government 2017).[8]

The Leasehold Reform (Ground Rent) Act 2022 also came into force in June 2022. The new legislation is the first step in the government reform package that aims to create a fairer housing system and levelling-up opportunities for more people. The Act means that any ground rent demanded as part of a new regulated residential long lease where a premium is paid, may not exceed more than one peppercorn per year. Most new leaseholders will not be faced with financial demands for ground rent. The Act also bans landlords from charging administration fees for collecting a peppercorn rent. If a landlord charges ground rent in contravention of the Act, they are liable to receive a financial penalty between £500 to £30,000 (Department for Levelling Up, Housing and Communities 2022).

In the context of this complex legal/policy landscape of housing in the UK, in Chapter 4 we will look at the specific housing issues that the Housing Ombudsman and the Property Chamber are able to help with. We now turn to the SEND context.

SEND

The wider context: the challenges in the system

The system to support children and young people with SEND has been fraught with challenges over the years. This has been compounded by a complicated redress system that leaves parents and carers unclear of where to go to resolve their dispute with the LA.

A child or young person with SEND is entitled to special education provision. According to the Special Educational Needs and Disability Code of Practice (2014), a child or young person (aged 0–25) can be categorized as having SEND, if they:

> have a significantly greater difficulty in learning than the majority of others of the same age; or has a disability which prevents or hinders him or her from making use of facilities of a kind generally provided for others of the same age in mainstream schools or mainstream post-16 institutions.[9]

A child/young person with SEND can qualify for extra support to enable them to participate in mainstream education within either a mainstream or specialist school, depending on the severity of their needs. In 2021, 1.4 million school pupils were identified with special educational needs making up 15.8% of all school children. Those identified as requiring SEND support made up 12.2% of all school pupils (an increase from 11.6% in 2016). The most common type of need in primary school in 2021 was related to speech, language and communication, and in secondary schools it was social, emotional and mental health (Department of Health and Social Care (DHSC) 2022: 7). To receive extra support, a child/young person with SEND needs to receive an education, health and care (EHC) plan from their LA to identify their educational, health and social needs and to set out what additional support is required to meet those requirements. An EHC plan is a legal document which identifies:

- a child's special educational needs;
- the additional or specialist provision (support, therapy, and so on) required to meet their needs;
- the outcomes (capabilities, achievements) the provision should help them to achieve; and
- the placement (the school or college they should attend).

The Children and Families Act 2014 Code of Practice 2015 provides guidance to LAs about how they should: assess children and young people for an EHC plan; decide whether to issue a plan; decide on the content of the plan; and implement, monitor or cease a plan (LGSCO 2019: 3). Then, after applying, the applicant's LA has 16 weeks to decide whether an EHC plan will be provided for the child and 20 weeks from initial application to create the final plan. Should a plan be offered by the LA, a draft plan will be created and sent to the applicant, who is able to comment on the contents including whether a request for their child to attend a specialist school has been granted; then the final plan is created.

However, while this process might seem straightforward, in practice, there have been significant challenges in children/young people being granted SEND support by their LA: for example, by having their application for an initial assessment rejected; or, even if an assessment has taken place, a decision can be made that no plan is required. Despite 93,302 plans being requested in 2020, only 62,180 were granted (GovUK 2002). This has resulted in growing frustration among parents/carers whose children have been left without the support they need to thrive in school.

Another challenge is the length of time it takes to process an EHC plan. The LA should take 20 weeks to deliver the plan but, in reality, it takes much longer – only 59.9% were issued within 20 weeks in 2021, leaving families waiting for long periods without the additional support for their child.

Further areas of contention are around alternative provision, where a young person's needs mean they cannot receive education in a mainstream setting. Parents/carers can specify in the plan which school they would like their child to attend, but they are not always given the school they have requested, let alone a place in a specialist school. If a parent/carer does not agree with the LA's decision to issue an assessment plan, or specialist school place, they can appeal to the First-Tier (Special Educational Needs and Disability) Tribunal.

Numerous reports have highlighted discontent from parents and carers. For example, the National Audit Office published a report in September 2019 entitled 'Support for pupils with special educational needs and disabilities in England' (National Audit Office 2019). It concluded that, while some pupils with SEND are receiving high-quality support, other pupils are not being supported effectively, particularly if they have not been granted EHC plans. Further, the Children's Commissioner's report (Children's Commissioner 2022: 5) highlighted that children with special educational needs or disabilities are a broad and diverse group and that the SEND system has to work for *all* children. The Commissioner also emphasized the challenges many children face to secure their support packages, and the importance that the support promised is delivered seamlessly.

Dissatisfaction with regards to children and young people's EHC plans was further demonstrated in a survey conducted during the pandemic, where 68% of parents reported that their child's needs were 'not met at all' or only 'somewhat met' in relation to their EHC plan (Ashworth et al 2022).

While there are considerable challenges relating to the provision of SEND services to children and young people who need additional support, the pressure and restraints on LAs should not be ignored; indeed, needs have outstripped funding thus making the system financially unsustainable. Two-thirds of LAs have been left with growing deficits, and by the end of 2020/2021 the national deficit was over £1 billion. Financial sustainability is impacted by increased demand for special school places, the increased use of independent schools and the reductions in per-pupil funding. In addition, there is a lack of consistency in the expenditure on different types of specialist provision for children and young people with SEND, costing more than double in an independent specialist school compared to an academy special school (Department for Education (DfE) 2021a, 2021b).

The lack of provision has led to a huge increase in the appeal rate to the First-Tier Special Educational Needs and Disabilities Tribunal, which has increased year on year since 2015. In 2020/2021, 8,600 SEN appeals were registered, an increase of 8% from the previous year. More significantly, 96% of cases at the Tribunal were overturned (on at least some parts of the appeal). This was an increase of 2% from the previous year. Similarly, the LGSCO reported increases of 45% from 2016–2017 to 2018–2019, and they upheld nearly nine out of ten investigations in 2018 (LGSCO 2019: 3). From January to March 2021, the PHSO received 8,258 enquiries, which was 12.4% more than for the same period in the previous year; and the PHSO completed 185 investigations, 41 fewer compared to the same three months a year earlier.[10] This increase, coupled with the high success rate for parents/carers, demonstrates the need for improvement to the decisions made on SEND provision by LAs.

Having a successful outcome at the tribunal or ombuds, however, comes at a cost to both the appellant and the tribunal/ombuds. Challenging a decision has a huge emotional impact on the parents/carers, and subjects them and their child to further delays. In the case of appeals, the burden to make the correct decision on a child/young person's SEND provision rests with the tribunal which also covers the costs of running the hearings. However, having an effective and seamless appeals process is vital in helping families to obtain the specialist support required. From the families' perspective, the experience of the SEND system is too bureaucratic and adversarial. There are too many difficulties and delays in securing support for children/young people, causing frustration to their parents and carers. Overall, navigating the SEND system is complex, requiring families to engage with multiple services and assessments, and the appeals process causes further delay, frustration and confusion about where to go for help.

In the Children Commissioner's investigation into the SEND system in 2022, she reported: 'Parents told me about the challenge of accessing the right help, quickly, for their children; having to navigate a complicated system that is too often adversarial. But they also told me how access to the right help has transformed their lives and their outlook' (Children's Commissioner 2022: 5).

Policy developments

In September 2014, significant reforms were undertaken in the SEND system under the Children and Families Act 2014. The aims of the reforms were for: children's needs to be identified earlier; families to be more involved in decisions affecting them; education, health and social care services to be better integrated; and support to remain in place (where appropriate) for children and young people from 0 to 25 years. EHC plans were introduced as a replacement for the statement of special educational needs. Co-production, joint working and a 0–25 child-centred approach were also introduced as part of the reforms.

While the reforms themselves were widely welcomed, their ambitions are yet to be realized, with a SEND system that is still not operating effectively and 'too many children and young people not fulfilling their potential, parental confidence in decline and further pressure on a system already under strain' (DHSC 2022: 8).

Under part 3 of the Children and Families Act 2014, the policy for children and young people with special educational needs was set out. It included provisions for LAs to keep under review the education, training and social care training provisions for children and young people aged 0–25 years who have a SEND. In relation to provision, the Act states that:

> The local authority must secure an EHC plan needs assessment for the child or young person if, after having regard to any views expressed and evidence submitted under subsection 7, the authority is of the opinion that
>
> (a) the child or young person has or may have special educational needs, and
> (b) it may be necessary for special educational provision to be made for the child or young person in accordance with an EHC plan. (Children and Families Act 2014, s 36(8))

In response to the increasing issues with the system, the Education Select Committee published a report on SEND in 2019. In the report they stated that the ambitions in the reforms remain to be realized: 'Let down by failures of implementation, the 2014 reforms have resulted in confusion and at times unlawful practice, bureaucratic nightmares, buck-passing and a lack of accountability, strained resources and adversarial experiences, and ultimately dashed the hopes of many' (House of Commons Education Committee 2019: 3).

This followed some additional improvements made by the government in 2018. In November 2018, the DfE published successful applications from trusts to LAs to run special free schools; and in March 2019, there was an announcement of 37 further successful LA bids for special schools. In addition, in December 2018, £350 million was allocated for high needs funding, which included money to increase the number of educational psychologists. In March 2019, an additional £31.6 million was announced for the operating costs of training providers (House of Commons Education Committee 2019: 9).

However, these improvements were not sufficient, and in 2019, after the realization that the 2014 reforms had not been successful, the SEND Review was launched jointly by the DfE and the Department for Health and Social Care (DHSC 2022). This was amid the growing concerns about the challenges facing the SEND system in England. The SEND Review focused on how the system had evolved since 2014 and looked at ways to improve it, to ensure that it could be more effective and that resources would be sustainable.

A green paper (DHSC 2022)[11] identified three key challenges facing the SEND system:

1. outcomes for children and young people with SEN or in alternative provision are poor;
2. navigating the SEND system and alternative provision is not a positive experience for children, young people and their families; and
3. despite unprecedented investment, the system is not delivering value for children, young people and families (DHSC 2022: 10–11).

The proposals set out in the green paper to mitigate these challenges were for:

• a single national SEND and alternative provision system;
• provision from early years to adulthood (including an increase to the schools budget by £7 billion by 2024–2025);
• a reformed and integrated role of alternative provision;
• system roles and accountabilities and funding reform; and
• delivering change for children and families (DHSC 2022: 14–17).

The consultation ran until July 2022, and a national SEND delivery plan was published in 2023 (HM Government 2023) setting out how the changes will be implemented. Since the publication of the green paper, there have been a number of criticisms on the proposals raised by stakeholders. For instance, on 21 November 2022, a group of 34 lawyers representing children and young people wrote to the Secretary of State for Education expressing the concern that the green paper risked diluting children and young people's rights to provision and the support that meets their individual needs. Fears were expressed that children and young people are not currently receiving the support they need. The letter referred to what the group of lawyers considered was unlawful decision-making by LAs. It was felt that the implementation of the 2014 reforms had been inadequate and that LAs had failed to fulfil their legal duties as set out in the legislation and regulations. The letter's authors also criticized the reduction in the choice of education setting and the requirement that families participate in mandatory mediation.[12]

Despite the criticism, the DfE has acknowledged that the SEND system requires significant improvement, and it is hoped that the white paper will have a positive impact on the provision of SEND to those who need it. The quality of decision-making across LAs has been acknowledged by both the DfE and the Ministry of Justice, who sit (as observers) on the Administrative Justice Council's[13] working group 'Improving first instance decision-making by local authorities', which seeks to understand the high overturn rate in tribunals and to provide practical solutions to improve decisions made by LAs (HM Government 2023).

Effects of the pandemic

The Ofsted 2021 Report found that existing weaknesses in the SEND system have been exacerbated by the pandemic, as children are more likely to have been 'out of sight' of services. These are, for example, weaknesses in universal education and in health and care services, resulting in children and young people not learning essential skills and knowledge, and mistakenly being identified as having SEND. The significant inconsistencies in how SEND is identified, a lack of joined-up commissioning and joint working across education, health and care has become ever more apparent. Further, there is a lack of clarity between organizations about who is responsible and accountable within local area SEND systems. This means delays in decisions and children not being able to receive the support they need.

The report makes several recommendations for improvement in the SEND system:

- more accessible universal services for children and their families, delivered by practitioners with a strong understanding of how to meet the needs of children and young people with SEND;
- more accurate identification when children need targeted or specialist support and higher aspirations for children and young people with SEND; and
- a greater sense of joint responsibility between partners in a local area, clearer accountability for different organisations within local systems, and greater coordination of universal, targeted and specialist local services so children get the right support at the right time. (Ofsted 2021)

As mentioned earlier, if parents/carers need to challenge a school or LA's decisions, there are different routes available. We look closely at those pathways through the tribunal and the ombuds in Chapters 4 and 5. Next, we turn to the help-seeker journey.

Mapping pathways to justice: the help-seeker journey

As a first step to navigating these processes, we have mapped help-seeker journeys in the areas of housing and SEND, and we have produced resources including a map, an animation, booklets and a report detailing the help-seeker journey.[14] The process of putting these resources together involved assembling information from various sources, including documents (previous research, statistics and reports from government, representative organizations, charities, relevant tribunals, LAs) and expert opinions, classifying and sorting the data into something that can be stored and used by help-seekers and practitioners alike. Any gaps in the data/information were filled by contacting relevant stakeholders about types of problems that the SEND Tribunal, the Property Chamber, the Housing Ombudsman the LGSCO and the PHSO encompass.

Our map was inspired by the Australian legal help journey map 'Joining up justice',[15] which followed a similar process of 'stages' for legal issues in Australia. The Justice Connect map is based on five years of extensive academic research evaluating the experience of people in Australia looking for legal help, and the experience of legal services delivering help. The researchers spoke directly with hundreds of legal help-seekers and held workshops with many community legal centres, legal aid providers and referring agencies. Overall, the research showed that, for help-seekers, looking for legal help is difficult, confusing and demoralizing. And for service providers, connecting with help-seekers, handling enquiries, getting the word out and reducing the referral roundabout is a constant challenge. Their interactive map summarizes the experience of hundreds of legal help-seekers, community legal centres, legal aid providers and referring agencies.

We adapted the Australian map's five steps of awareness, consideration, engage, service and outcome to create an eight-step help-seeker journey consisting of:

1. awareness
2. taking action
3. advice sector referral, support and guidance
4. intermediate processes
5. consideration
6. engage
7. service
8. outcome (Figure 3.1).

It was necessary to adapt the Australian legal help journey map from a simpler 'problem, response, outcome' process of receiving legal advice which is linear, to a non-linear two-part process: the first maps getting advice, and the second maps the dispute resolution channel. Another difference from the Australian help-seeker journey is that advice is referred to as a 'service', which is not the case in the UK where we look at the use of advice, ombuds and tribunals (see Figure 3.1).

Figure 3.1: The eight steps in the help-seeker journey

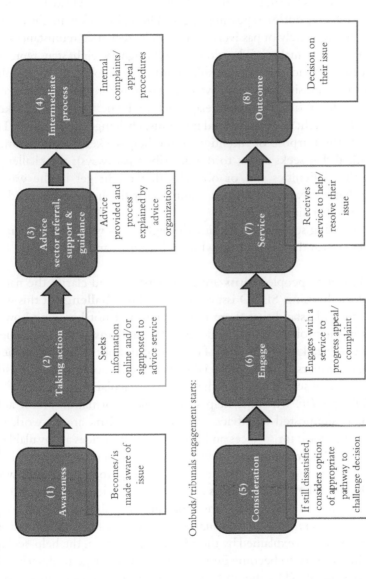

Note: Although this representation is linear, it is not as straightforward in reality. The help-seeker may navigate the process differently, missing out or repeating stages, and often multiple things happen alongside each other. The help-seeker might abandon the journey and/or get stuck along the way. The help-seeker might engage with the process actively or passively, and the help-seeker's circumstances can affect their decision-making.

We also built on the Ministry of Justice visual representation of the steps and 'pain points' users go through when seeking legal advice. These eight steps represent the *ideal case* (as we will see in the following chapters, this is rather artificial and expects the help-seeker to be informed and educated) of a person going through the system to seek help for their problems. Although the representation is linear, in reality the help-seeker may navigate the process differently, missing out or repeating stages. The help-seeker might engage with the process actively or passively, and the help-seeker's circumstances can affect their decision-making. Therefore, even though the eight steps represent the ideal case scenario, the steps are still dauntingly complex, with a number of essential prerequisite steps that have to be taken before the next stage can be proceeded to, or time that has to elapse before the next step can be taken.

Each of these eight stages is divided into aims, challenges and actions. The aims set out the purpose or intention of the help-seeker (in our example the individual that seeks access to the described pathways); the challenges highlight the difficulties and/or obstacles that might get in the way of achieving that aim; and the actions outline what the help-seeker needs to do to achieve that aim.

In the following section, we outline the eight steps with their associated aims, challenges and actions upon which Chapters 4 and 5 are built.

Step 1. *Awareness*: people's **aims** are to understand and clarify the nature of the housing/SEND issue they face. The **challenge** at this stage is to understand that there is a problem but not knowing how to stop the problem from happening. The **actions** at this stage might include talking to friends, family, community networks, general practitioners, health and social workers, and searching online using a range of search terms in their own vernacular.

Step 2. *Taking action*: when the help-seeker seeks information online and/ or is signposted to advice services, their **aims** are to work out how to look for services and find out what services are available to them. The **challenges** include how and where to look for help, and knowing which service is the most appropriate to help with their problem. The **actions** are gaining knowledge of where to go for help and having the confidence in being able to do that.

Step 3. *Advice sector referral, support and guidance*: when advice is provided and the processes explained by the advice organization, the help-seeker's **aim** is to try to become empowered with language to describe the housing/SEND issue and to find out what relevant institutional options are available to help them with their housing/SEND issue. The **challenges** lie in identifying the legal dimension of a problem and knowing, or engaging with, the relevant terminology. Here the **actions** are to engage with advice services.

Step 4. *Intermediate processes: making a complaint*: when the help-seeker completes the internal complaints/appeal procedures, with or without support from the advice sector organization, they **aim** to understand the internal complaint process by taking their complaint to the LA or decision-making body. The **challenge** might be finding it difficult to proceed without appropriate support. The **actions** in this stage are to complete the internal complaint process of the LA or body complained about.

Step 5. *Consideration of alternative pathways*: if the help-seeker is still dissatisfied with the outcome of the internal processes (that is, if the complaint is not resolved), he/she considers options of appropriate pathways to challenge the decision. The help-seeker's **aim** is to understand the relevant institutional options available and the differences between the options and their limits and then to choose an institution to engage with. The **challenges** are to seek information about housing/SEND issues as they are often not specific enough for the help-seeker's situation, or information about housing/SEND issues may not be user-friendly nor accessible or up to date; institutions might use inconsistent language and terminology making it difficult to compare them; institutions might also not have their eligibility criteria available online, making it difficult to assess which service is most appropriate for the help-seeker's situation. The **actions** in this stage are understanding and evaluating avenues for progressing the housing/SEND issues and comparing and evaluating available institutions within those avenues.

Step 6. *Engage*: when the help-seeker engages with a redress mechanism to progress an appeal/complaint, the help-seeker's **aims** are to find the contact details of the redress mechanism or appealing/engaging through the advice sector organization; to engage with the right entry point or application process easily; to find out quickly if their complaint is eligible; and to get guidance on where else to look for help. The **challenges** might be inconsistent information about service eligibility, or no information about eligibility guidelines; help-seekers must wait for intake assessments to take place to be advised of whether they can be assisted, and, for some, they will be told they are ineligible after spending significant time on the intake process; online application forms can be lengthy and difficult to understand or complete (for those who are less digitally literate). The **actions** include understanding the institution's complaint's process and completing the application online or offline.

Step 7. *Service*: when the help-seeker receives service to help resolve their housing/SEND issue, they **aim** to understand how institution/

redress mechanisms will progress the issue and what next steps to take. The **challenges** are that redress mechanisms/institutions are slow, and many cannot assist with urgent requests. The **actions** in this stage are to receive direct assistance.

Step 8.   *Outcome*: when the help-seeker receives a decision on their housing/ SEND issue, the help-seeker **aims** to resolve the issues with the best possible outcome. Challenges might be that the outcome did not meet the help-seeker's expectations or improve the help-seeker's situation; or the help-seeker does not understand the outcome. **Actions** in this stage are receiving the outcome from the organization.

In sum, the aims, challenges and actions apply to the help-seeker journey in the areas of both housing and SEND. However, the steps to redress differ by area. We address these in detail in Chapters 4 and 5. Importantly, our map of the pathways to seek help for housing and SEND issues shows the ideal case of how advice and justice can be accessed. In reality, it is not straightforward; most people do not know how to access these pathways, which leaves the system (more) accessible for those who are savvy and can navigate it. The complex and siloed system leaves the help-seeker in a vulnerable position if they are not able to navigate it. In our discussion of the housing and SEND pathways in the following chapters, we draw on interviews with members of the public to highlight how those people who go through the process experience it.

Notes

[1]   See Administrative Justice Council Website (2023). https://ajc-justice.co.uk/

[2]   The Channel 4 programme *Untold: Help! My Home is Disgusting* illustrates the scale of the National Housing Crisis. www.channel4.com/programmes/help-my-home-is-disgusting-untold/on-demand/74288-001

[3]   See also Office for National Statistics (2022) for most recent property prices, private rent and household statistics.

[4]   2019 Manifesto Pledge. www.conservatives.com/our-plan

[5]   'Policy Paper: A Fairer Private Rented Sector', chapter 4. www.gov.uk/government/publications/a-fairer-private-rented-sector/a-fairer-private-rented-sector#chapter-4-improved-dispute-resolution

[6]   'Consultation outcome: call for evidence to consider the case for a Housing Court: government response' (16 June 2022). www.gov.uk/government/consultations/considering-the-case-for-a-housing-court-call-for-evidence/outcome/call-for-evidence-to-consider-the-case-for-a-housing-court-government-response

[7]   https://bills.parliament.uk/bills/3177

[8]   See also the Institute for Government's Policy Tracker that has tracked the progress of major government legislation, policies and projects and the impact of COVID-19 on the government's programme of work in the area of housing and homelessness: 'Trackers'. www.instituteforgovernment.org.uk/policy-tracker#housing

[9]   'Statutory Guidance: SEND code of practice: 0 to 25 years' (2014), xiv, 16. www.gov.uk/government/publications/send-code-of-practice-0-to-25

[10] PHSO website, 'January to March 2021 performance statistics'. www.ombudsman.org. uk/about-us/corporate-information/how-we-are-performing/performance-statistics/ january-march-2021-performance-statistics

[11] See also SEND and AP Green Paper: Responding to the consultation (29 March 2022). www.gov.uk/government/publications/send-and-ap-green-paper-responding-to-the-consultation

[12] 'Letter to Secretary of State for Education from lawyers who represent children and young people with special educational needs & disabilities (SEND)' sent by IPSEA (Independent Provider of Special Education Advice). www.ipsea.org.uk/Handlers/Downl oad.ashx?IDMF=e031d3f6-bf84-47f0-b622-f385b3627230

[13] The Administrative Justice Council is the oversight body to the administrative justice system across the UK. www.ajc-justice.co.uk

[14] Delivering Administrative Justice, 'Delivering administrative justice after the pandemic: what can we learn about digitalisation and vulnerable groups?' www.ucl.ac.uk/jill-dando-institute/research/centre-global-city-policing/cgcp-research/cgcp-delivering-administrat ive-justice and Nuffield Foundation: 'Delivering administrative justice after the pandemic'. www.nuffieldfoundation.org/project/delivering-administrative-justice-after-the-pandemic

[15] Legal Help Journey Map. https://joiningupjustice.org.au

References

Ashworth, E., Kirby, J., Bray, L. and Alghrani, A. (2022) 'The impact of the COVID-19 pandemic on the education, health and social care provision for children with Special Educational Needs and Disabilities (SEND): The Ask, Listen, Act Study', Liverpool: Edge Hill University, Liverpool John Moores University, University of Liverpool and Liverpool Health Partners. www.ljmu.ac.uk/~/media/files/ljmu/research/centres-and-institutes/ rcbb/quantitative-evidence-briefing.pdf?la=en

Children's Commissioner (2022) 'Beyond the labels: a SEND system which works for every child, every time'. www.childrenscommissioner.gov.uk/ report/a-send-system-which-works-for-every-child-every-time/

Creutzfeldt, N., Gill, C., Cornelis, M. and McPherson, R. (2021)) *Access to Justice for Vulnerable and Energy-Poor Consumers: Just Energy?* London: Bloomsbury.

Crisis (2022) 'The Homelessness Monitor: England 2022'. www.crisis.org. uk/ending-homelessness/homelessness-knowledge-hub/homelessness-monitor/england/the-homelessness-monitor-england-2022/

Department for Education (2021a) 'LA and school expenditure'. https:// explore-education-statistics.service.gov.uk/find-statistics/la-and-school expenditure/2020-21

Department for Education (2021b) 'Schools, pupils and their characteristics'. https://explore-education-statistics.service.gov.uk/find-statistics/school-pupils-and-their-characteristics

Department of Health and Social Care (2022) 'SEND Review: right support, right place, right time', London: HM Stationery Office. https://assets.pub lishing.service.gov.uk/government/uploads/system/uploads/attachment_ data/file/1063620/SEND_review_right_support_right_place_right_time _accessible.pdf

Department for Levelling Up, Housing and Communities (2022) 'Leasehold Reform (Ground Rent) Act 2022: guidance for leaseholders, landlords and managing agents'. www.gov.uk/government/publications/the-leaseh old-reform-ground-rent-act-user-guidance/leasehold-reform-ground-rent-act-2022-guidance-for-leaseholders-landlords-and-managing-agents

Gill, C. and Creutzfeldt, N. (2018) 'The "ombuds watchers": collective dissent and legal protest amongst users of public services ombuds', *Social and Legal Studies* 27(3): 367–388.

Gibson, J.L. and Caldeira, G.A. (1996) 'The legal cultures of Europe', *Law & Society Review* 30(1): 55–85.

GovUK (2002) 'Reporting year 2022: education, health and care plans'. https://explore-education-statistics.service.gov.uk/find-statistics/educat ion-health-and-care-plans#dataBlock-324a7b1c-4388-4a54-a1b2-b4b1b 5125a3b-charts

Hazel, G. and Beinart, S. (1999) *Paths to Justice: What People Do and Think about Going to Law*, Amsterdam: Criminal Justice Press.

Hertogh, M. (2009) 'What's in a handshake? Legal equality and legal consciousness in the Netherlands', *Social and Legal Studies* 18(2): 221–239.

Hilber, C. and Schoni, O. (2016) *Housing Policies in the United Kingdom, Switzerland, and the United States: Lessons Learned*, Tokyo: ADBI Institute.

HM Government (2023) 'Special Educational Needs and Disabilities (SEND) and Alternative Provision (AP) Improvement Plan: right support, right place, right time', March 2023. https://assets.publishing.service.gov.uk/gov ernment/uploads/system/uploads/attachment_data/file/1139561/SEND_ and_alternative_provision_improvement_plan.pdf https://www.adb.org/ sites/default/files/publication/183139/adbi-wp569.pdf

House of Commons (2022) 'Social housing reform in England: what next?' https://commonslibrary.parliament.uk/research-briefings/cbp-9227/

House of Commons Education Committee (2019) 'Special Educational Needs and Disabilities', first report of session 2019. https://publications. parliament.uk/pa/cm201919/cmselect/cmeduc/20/20.pdf

LEF (2020) 'Digital justice: HMCTS data strategy and delivering access to justice report and recommendations'. https://research.thelegaleducatio nfoundation.org/wp-content/uploads/2019/09/DigitalJusticeFINAL.pdf

Local Government Association (2017) 'Homelessness Reduction Act (get in on the act)'. London: Local Government Association. www.local.gov. uk/publications/homelessness-reduction-act-get-act

Local Government and Social Care Ombudsman (2019) 'Not going to plan? Education, health and care plans two years on', Coventry: LGSCO. www. lgo.org.uk/information-centre/news/2019/oct/a-system-in-crisis-ombuds man-complaints-about-special-educational-needs-at-alarming-level

LSE (2019) 'The UK's housing crisis: what should the next government do?' https://blogs.lse.ac.uk/politicsandpolicy/housing-crisis-what-should-the-next-government-do/

Ministry of Housing, Communities and Local Government (2017) 'Homelessness Reduction Act 2017: government response to the call for evidence'. www.gov.uk/government/consultations/homelessness-reduction-act-2017-call-for-evidence/outcome/homelessness-reduction-act-2017-government-response-to-the-call-for-evidence

Ministry of Housing, Communities and Local Government (2018a) 'A new deal for social housing'. www.gov.uk/government/consultations/a-new-deal-for-social-housing

Ministry of Housing, Communities and Local Government (2018b). 'Review of social housing regulation: call for evidence'. www.gov.uk/government/consultations/review-of-social-housing-regulation-call-for-evidence

Ministry of Housing, Communities and Local Government (2020) 'The Charter for Social Housing Residents: social housing white paper'. www.gov.uk/government/publications/the-charter-for-social-housing-residents-social-housing-white-paper

National Audit Office (2017) 'Housing in England: overview'. www.nao.org.uk/reports/housing-in-england-overview/

National Audit Office (2019) 'Support for pupils with special educational needs and disabilities in England'. www.nao.org.uk/reports/support-for-pupils-with-special-educational-needs-and-disabilities/

Office for National Statistics (2022) 'Housing: property price, private rent and household statistics'. www.ons.gov.uk/peoplepopulationandcommunity/housing

Ofsted (2021) 'Research and analysis: SEND: old issues, new issues, next steps'. www.gov.uk/government/news/children-and-young-people-with-send-disproportionately-affected-by-pandemic

Organisation for Economic Co-operation and Development (2020) 'Housing amid Covid-19: policy responses and challenges'. www.oecd.org/coronavirus/policy-responses/housing-amid-covid-19-policy-responses-and-challenges-cfdc08a8/

Schofield, M. (2021) 'Fobbed off: the barriers preventing women accessing housing and homelessness support, and the women-centred approach needed to overcome them', London: Shelter.

Shelter (2018) 'Building for our future: a vision for social housing'. https://england.shelter.org.uk/support_us/campaigns/a_vision_for_social_housing

Tomlinson, J., Hynes, J., Maxwell, J. and Marshall, E. (2020) 'Patterns in judicial review during the COVID-19 pandemic'. https://adminlawblog.org/2020/05/28/joe-tomlinson-jo-hynes-jack-maxwell-and-emma-marshall-judicial-review-during-the-covid-19-pandemic-part-ii/

4

Pathways Through the
AJS: Housing

Introduction

This chapter explores the pathways to redress for housing problems available to people guided by (the ideal case) help-seeker journeys. This helps us to understand how access points have been compromised and which pathways to justice are difficult to negotiate or blocked (Genn 1999; McKeever et al 2018). In this context, we look at the Housing Ombudsman and the Property Chamber that provide redress for housing problems. These AJS institutions do not typically interact with one another, although some progress was made during the pandemic to develop a relationship.

In this chapter we build on the outline of the legal/policy context of housing in England. First, we present the ideal case help-seeker journey for those in need of support for housing problems using the map we have developed. Then, we will draw on interviews with advice sector professionals, judges, case handlers and users to show how the help-seeker journey unfolds in reality.

Pathways to resolve grievances: ombuds and tribunals

Housing problems can be varied and complex. The scope of our project limits these to the housing issues that the Housing Ombudsman and the Property Chamber deal with. They can relate to residential property, land registration and agricultural and drainage matters. Common housing issues include but are not limited to: residential property: repairs and tenant behaviour; land registration matters – disputes over a change to the land register; agricultural land and drainage matters – disputes between agricultural tenants and landlords in relation to certain farming tendencies.

An area of overlap in jurisdiction to mention here briefly are the County Courts. They are currently being reformed and have a similar jurisdiction to the Property Chamber. The County Courts deal with civil (non-criminal) matters. It is important that users, who are similar to the users of tribunals, do not have differing journeys. As a matter of policy, it may be that the way in which County Courts deal with possession cases will change, perhaps within the next two years, because the 'Renters Reform Bill' (Department for Levelling Up, Housing and Communities 2022) is about to be introduced. According to Michael Gove (Secretary of State for Levelling Up, Housing and Communities and Minister for Intergovernmental Relations):

> For too long many private renters have been at the mercy of unscrupulous landlords who fail to repair homes and let families live in damp, unsafe and cold properties, with the threat of unfair 'no fault' evictions orders hanging over them. Our New Deal for renters will help to end this injustice by improving the rights and conditions for millions of renters as we level up across the country and deliver on the people's priorities.

If this reform happens, then it will create an opportunity to revisit the ways in which housing disputes are dealt with. In addition, one of the consequences of this Bill would be a huge increase in the number of rent cases.

Further, the most recent death of a toddler (Brown and Booth 2022) due to black mould in the social housing flat where the child's family were living, provides some insight into the scale of the problem in the social housing sector in England (Housing Ombudsman 2023).

Complaints/appeals about residential property, land registration and agricultural land and drainage

On the surface it appears that there are overlaps in the kinds of housing issues the Housing Ombudsman and the Property Chamber handle. Similar types of housing issues might be dealt with by either the Housing Ombudsman or the Property Chamber. The Housing Ombudsman will consider complaints concerning matters that they can deal with more quickly, more fairly, more reasonably, or more effectively than were the complainant to seek a remedy through the courts, a designated person, or other tribunal or procedure, such as the Property Chamber. Moreover, the Property Chamber (but not the Housing Ombudsman) handles applications, appeals and references relating to disputes over property (for example, appeals against legal notices, banning orders and rogue landlords, and charges).

Indeed, upon closer inspection it becomes clear that the problem areas that both institutions deal with are distinct. The Housing Ombudsman is predominately concerned with residential property issues, where the most common problem area is repairs, followed by tenant behaviour.

The Property Chamber, on the other hand, mainly handles disputes between lessors and lessees, appropriate levels of rent payable by tenants, action to ensure compliance by landlords with various obligations within the jurisdiction of the tribunal, and registration and agricultural land and drainage matters. However, there are two significant areas which have been identified by the Property Chamber and Housing Ombudsman as having a sufficient level of cross-over – namely, service charges and rent arrears. There are, though, distinct differences in the aspects of these claims that each institution can look at. For service charges, the Ombudsman's jurisdiction looks at the process around the service charge rather than the level of the charge; the amount of service charge is a matter for the tribunal. This is in accordance with paragraph 39(g) of the Housing Ombudsman Scheme which states that they cannot consider complaints which 'concern the level of rent or service charge or the amount of the rent or service charge increase'.

Information about what areas are outside of the jurisdiction is explained in the ombuds reports of the investigations. In the Guinness Housing Association Limited complaint in June 2022, regarding the landlord's handling of the resident's rent and service charge payments, the scope of the investigation was made clear in the report:

> It is important to be aware that it is outside the role of the ombudsman to review complaints about the increase of service charges and determine whether service charges are reasonable or payable. However, we can review complaints that relate to how information about service charges was communicated by the landlord. This is in line with paragraph 39(g) of the Housing Ombudsman Scheme, which states that the ombudsman will not consider complaints that concern the level of service charge or rent or the increase of service charge or rent. Complaints that relate to the level, reasonableness, or liability to pay rent or variable service charges are within the jurisdiction of the First–Tier Tribunal (Property Chamber) and the resident would be advised to seek free and independent legal advice from the Leasehold Advisory Service (LEASE)[1] in relation to how to proceed with a case if she wishes to pursue this aspect of his complaint further.[2]

The Property Chamber and Housing Ombudsman fostered a strong working relationship during the pandemic, with signposting and training between the two institutions. The Ombudsman, for example, will signpost cases to the tribunal if the complaint is outside its jurisdiction, or if most elements can be dealt with by the tribunal (where it would make sense to deal with the whole case in one place).

Complaints/appeals process: Housing Ombudsman and the Property Chamber

In the following section we provide the details of how to bring a complaint to the Housing Ombudsman and the Property Chamber, respectively, as per information available on their websites.

Making a complaint to the Housing Ombudsman

The steps to take when bringing a complaint to the Housing Ombudsman are divided into the following stages: tell the landlord about the problem; complain to a designated person; escalate the complaint to the Ombudsman; and the consideration stage.[3]

1. *Tell the landlord about the problem*[4]
   The first step for complainants is to report the problem to their landlord who may be able to put things right.[5] If the complainant has difficulty reporting the issue or is dissatisfied with the service they receive in response, the Housing Ombudsman can help the complainant and their landlord resolve the issue. All landlords have complaint's procedures that should be easy to use, fair and designed to put things right. If the complainant thinks their complaint is not being dealt with correctly, for example if they receive a delayed or no response, the Housing Ombudsman can help ensure their complaint is responded to by their landlord.

2. *Complain to a designated person*
   If the complainant is unable to resolve their complaint through their landlord's complaints procedure, they can contact a designated person who can also help find a solution. The designated person can be a Member of Parliament (MP), a local councillor or a tenant panel. Their role is to help resolve disputes between tenants and their landlords which they can accomplish in whatever way they think is most likely to work. If the designated person cannot help, they can refer the complaint to the Ombudsman. If the complainant has decided not to contact a designated person, they can go directly to the Ombudsman eight weeks after their landlord has given them a final response to their complaint.

3. *Escalate your complaint to the Ombudsman*[6]
   Before the individual makes a complaint, they will need to answer a few questions online. The complainant will then be taken to the online complaint form or signposted to other helpful information.[7]

4. *Consideration of the complaint*
   The Housing Ombudsman will deal with each complaint to find the best outcome for the specific circumstances involved. Once the Ombudsman receives the complaint they may:

- refer the case to a different organization if it is an issue they cannot make a decision about because it is not in their jurisdiction;
- work with the complainant and their landlord to resolve the dispute under their early resolution procedure – for example, the ombudsman can use their experience of resolving complaints to make suggestions to the landlord and/or the resident if they believe there is a way to resolve the complaint;[8]
- carry out an investigation: the ombudsman service only does this for those complaints where they decide an investigation is proportionate to the circumstances and evidence before them, for example complex complaints involving many issues.[9]

Applying to the Property Chamber involves a different set of steps. We outline these in the next section.

Applying to the Property Chamber

1. *Get help and advice before you apply*
   The applicant should contact Leasehold Advisory Service or Citizens Advice or another advice sector organization. The complainant can also get legal advice, including from a lawyer.
2. *Consider mediation*
   The applicant should consider whether they can use other methods of settling the dispute, such as ADR.[10] Mediation is a way of resolving disputes through effective communication and compromise. Mediation involves a third party[11] acting as a go-between to ensure the parties are able to communicate with one another.
3. *Apply to the Property Chamber*
   To apply to the tribunal, the applicant will need to fill in an application form.[12] Forms can also be obtained from a regional rent assessment panel. If no specific form exists for the applicant's case category, then the applicant should write to the tribunal including specified information. There is a fixed fee of £100 for most applications to the tribunal (it increases depending on the nature of the hearing identified as required to dispense with the issue). There are arrangements in place for the fee not to be charged in some circumstances, for example if the applicant is receiving certain benefits.
4. *Consideration of your application*
   Once an application is received, it will be assessed to check that the form is correctly completed and that the required attachments are present. If something is missing the Property Chamber will request this from the applicant and the application will not be accepted until all the required information and attachments are provided. If the required information is not provided, the application will not be accepted by the Property Chamber and the case will be closed. The applicant will be advised of

this, and informed that they may submit a fresh application when they have all the required information and documents. Complete applications will be passed to the member with delegated powers within the Property Chamber, who will consider the application. The Property Chamber will decide how best to progress the case. The Property Chamber will write to the applicant and any other parties to notify them of what will happen next (including information on the hearing).[13]

In the next section we present the ideal case help–seeker journey for those in need of support for housing problems, using the map we developed. We then discuss real–life case studies of the housing pathways from Step 1 (awareness) to Step 8 (outcome).

The ideal case help–seeker journey: a fictional case study

Here we outline the ideal case help–seeker journey for those in need of support for housing problems, using the map we developed, and provide a case study of the housing pathways from Step 1 (awareness) to Step 8 (outcome). We have created this fictional scenario to help illustrate some of the problems that people have and what they are required to do to access the system to seek advice, help and resolution.

Marta's housing journey

Marta is 40 years old, a single mother of two living in social housing, who has been struggling to pay her housing costs since COVID-19 hit.

Problem: because Marta lives in social housing; her landlord is a housing association. She has experienced difficulties with paying her rent since COVID-19, causing her to lose her job. She was able to continue with most of her regular payments. She has now found a part-time job and been able to clear some of her arrears and pay the ongoing rent. However, her landlord has now given notice that the rent is to be increased and Marta cannot afford it.

Marta goes through the HOUSING pathways:

Step 1 (Awareness)
Marta wants to know what she can do about her problem. She asks her friends and searches online. She wants help to deal with the proposed rent increase by her landlord.

Step 2 (Taking action)
Marta comes across an advice sector organization and arranges an appointment with them to find out how they can help her.

Step 3 (Advice sector referral, support and guidance)
The advice organization provides Marta with advice and explains where she can apply for extra support.

Step 4 (Intermediate processes)
Marta also talks to her housing association landlord about her situation. They are very unhelpful. They tell her that they are short of staff and that she has no choice but to pay the increase to the rent. They say that, if she refuses to pay, they will consider repossessing her home. Marta raises a complaint to her landlord about how she was treated.

Now Marta has several options. An advice organization will assist Marta to choose the pathway that is best for her specific problem. We discuss two here: she can take her problem to the Property Chamber *or* go to the Housing Ombudsman.

First, we will accompany Marta to the Housing Ombudsman.

Step 5 (Consideration)
Marta feels that her complaint is not being dealt with correctly by her landlord. She goes to an advice organization for help. An adviser explains that the Housing Ombudsman can help to make sure that her complaint is dealt with properly.

Step 6 (Engage)
After having tried to solve her problem with the housing association, Marta needs to wait for eight weeks before she can complain to the Housing Ombudsman. She submits her form online to the Housing Ombudsman. Marta can also go through her MP, a local councillor or a tenant panel to reach the Ombudsman, but Marta decides to skip this step and go straight to the Ombudsman.

Step 7 (Service)
After the complaint handlers have checked Marta's complaint, the Housing Ombudsman works with Marta and her landlord to resolve the dispute. Marta waits for a decision from the ombudsman.

Step 8 (Outcome)
After a few months, the Housing Ombudsman makes a decision. They suggest that the landlord should resolve the issue. They encourage the landlord to give Marta longer to pay the arrears and to reduce the increase in

rent back to the original amount. The landlord complies with the decision. In this example, Marta has a successful outcome, and she is happy that she no longer needs to pay a higher rent.

The other option for Marta is to bring her complaint to the Property Chamber.

Step 5 (Consideration)
Marta's complaint to the housing association landlord is unsuccessful. So, the advice organization helps her figure out what to do next. She can take her problem to the Property Chamber which is a tribunal, but first she tries mediation to see if she can come to an agreement with her landlord without the need for a judge. A housing mediation service tries to help Marta and her landlord reach an agreement, but they are unsuccessful.

Step 6 (Engage)
Marta appeals to the Property Chamber for a decision about the proposed rent increase. Marta needs to fill in a form to make the appeal. She downloads the form from the 'HM Courts & Tribunals Service' website. She fills it in and posts it, although email was also an option. Usually there is a fee of £100 to pay, but there is a 'fee waiver' available for those who need it. The advice organization helps Marta get the fee waiver, so she does not have to pay the £100.

Step 7 (Service)
Marta waits to hear back from the Property Chamber. The Property Chamber checks Marta's form and the extra attachments she sent with it. It then gets back to Marta with a timetable for her case, the date of her hearing, and some extra information about the hearing. The hearing is arranged to take place by video. Marta is confident with computers, so she is happy with this. But she is told that she can go to the tribunal in person if she would prefer. Marta attends the hearing at the tribunal. A panel of one judge and one or two other people who know about housing issues takes the decision. An LA representative is also at the hearing.

Step 8 (Outcome)
The Property Chamber writes to (or emails) Marta to tell her their decision and sends a copy of their decision to the landlord. The Property Chamber decides that the increased rent is more than it would be for similar properties and that the increase would be unreasonable. It is decided her current rent is accurate and Marta therefore does not need to pay the increased rent.

Recall that at Step 5 Marta has several options to seek help. An advice provider will assist Marta in choosing the pathway that is best for her specific problem. We discussed two options: the Property Chamber and the Housing

Ombudsman. One of the differences between these two processes is that the outcome of a tribunal is legally binding, whereas that of the Ombudsman is a recommendation that the authority complained about can choose to comply with (anecdotally there is a high rate of compliance). We will explore the differences in duration, satisfaction and compliance with the processes in the next section (using our interview data). Specifically, we will next draw on interviews conducted with users to present how the help-seeker journey unfolds in reality.

The help-seeker journey in reality: reflections from users

Our representation of the help-seeker journey is linear and compacted: it is not as straightforward in reality. The help-seeker may navigate the process differently, missing out or repeating stages, and often multiple issues can happen alongside each other. In addition, the help-seeker might abandon the journey and or get stuck along the way. Moreover, the help-seeker might engage with the process actively or passively, and the help-seeker's circumstances can affect their decision-making.

The datasets we draw upon for this chapter are the Property Chamber and Housing Ombudsman user interview data. We include two exemplary case studies from interviewees, one for the Housing Ombudsman, and one for the Property Chamber. For the sake of illustrating that the help-seeker journey differs across cases, we make use of different user experiences.

Housing Ombudsman

In the following we share a Tenants and Residents Association Chair's journey to, and through, the Housing Ombudsman during the pandemic to illustrate the help-seeker journey.

Step 1 (Awareness)

The Tenants and Residents Association was formed as a response to the many problems residents faced with their landlord (PA Housing, that is, a housing association) – problems that only became worse when COVID-19 hit:

> 'Well, our landlord is PA Housing, a housing association. … We've had a lot of problems with service delivery, with communication, with their failure to deal with issues like handling social behaviour, problems like that. So when we formed, coming up on three years ago, we were in a kind of crisis situation. There was just no communication, there was massive breakdown in service delivery, and, you know, things were spiralling. … So we were formed out of a lot of tensions and problems. … And then when the epidemic

kicked in they shut everything down, so that's really the background and where we're at.'

The interviewee described his building's housing security problem as continuous, commencing a few years ago, and, although it has been dealt with intermittently over time, it still currently persists.

'Well, failures of contractors, basically. We've had a contract that, up until now, it's just about to now, has been 50% security and 50% cleaning, and the company has never ever provided security, we just don't have security. And a couple of years ago we had weed dealers in the stairwells, and a couple of residents were attacked, and one boy was actually stabbed on the edge of the estate and ran around the whole of the estate trailing with, like, gallons of blood. Luckily, he survived. But at that point we kind of, like, went mad, and insisted that there and then they bring in security, which they brought it in for a period of time, for a few months we had security, but then it was withdrawn again.'

Step 2 (Taking action)

Initially the residents took the problem into their own hands by collectively participating in a service charge strike against the organization involved (PA Housing):

'We actually had a service charge strike a few months ago as a protest for a few weeks, and I think there were 15 residents who got involved. Lease holders were scared, the mortgage companies, and other people are like, you know, "We won't win, what's the point?" But I thought 15, you know, out of 116 was quite a good protest.'

Other respondents in our data jumped Steps 2, 3 and 4 and went straight to considering where to take their complaint (Housing Ombudsman or Property Chamber).

Step 3 (Advice sector referral, support and guidance)

The residents then contacted the council about their problems, as well as a social housing charity to seek support and guidance:

'We tried to keep the council involved, especially around issues like anti-social behaviour, because they've got a panel that sits across the borough for all landlords. ... So we try to talk to whoever we can, we're involved, we're affiliated to XXX as well, you know XXX? They've got, kind of, links into, you know, government departments and stuff, they negotiate with government departments. ... So they kind of keep us updated, we feed them back what's going on here, that kind of stuff.'

The residents also often contact their local housing advice service:

'There's a local housing advice service who we went to when we first formed, and they recommended a solicitor to call. … They're involved in the government's advisory process around the last housing Bill. It's this process where you can make a claim against your landlord on the basis that you've got internal and external repairs that have not been carried out. Housing Disrepair Protocol, and you can do that as a no win, no fee, so that's why we were able to do it, because none of us have got any money. And although the landlord agreed to carry out an inspection of all properties and carry out the necessary repairs, they did the inspection and then didn't carry out the repairs, so it was partially successful.'

Step 4 (Intermediate processes)

The residents had complained to the organization involved prior to reaching this stage of the process. This case, therefore, illustrates that the help–seeker may navigate the process in different ways and might engage with the process actively or passively. It also shows that the help-seeker's circumstances can affect their decision–making. In this case, the help-seekers have joined forces (forming the Tenants and Residents Association), allowing them to engage with the process more actively than they would if they were acting alone. Respondents did not always complain to the organization involved and instead went directly to either the Housing Ombudsman or the Property Chamber.

Step 5 (Consideration)

Having failed to resolve their problems via the aforementioned routes, the residents escalated their complaint to the Ombudsman: *"and in the end we gave up … we've taken it to the ombudsman and we're still waiting for a reply from the ombudsman."*

At this point in the process, participants reported taking their problem either to the Property Chamber or the Housing Ombudsman.

Steps 6 and 7 (Engage and Service)

Reflecting on the complaints process undertaken by the Residents and Tenants Association (group of residents), the interviewee explained:

'I mean, one of the problems with the ombudsman is you can only complain as an individual, so I can't complain as chair of my TA, I have to complain as me, my name. And you have to have gone through your landlord's own process. They've got a two-stage complaints process, very often the landlord will say, "Well, I'm just not even taking your complaint, you're just wrong, we're not going to deal with it". Or they formally state, "We're not going to take it to the second stage", things like that. And so formally, officially, you're

not even allowed to go to the ombudsman then but, you know, if you then complain to the ombudsman, say, "Look, they're refusing to comply with their own procedure", sometimes, not every time, but sometimes the ombudsman will order them to comply with their own procedure. So, then you can start again.'

It is clear that the process is lengthy and frustrating for complainants:

'But it's a very frustrating process, it's very bureaucratic, it's very distant, you know. … So they're not very hands on, they just rely on what you send them and what the landlord sends them. They don't give you any feedback, they don't come back to you and ask you, "Can you explain this, and explain that?" You know? "It looks like you're saying this but you haven't got evidence for that", you know? There's no, like, interrogative process, which I feel there should be. You know, I'm just an individual, I'm not a housing professional. If they want to know something, you know, they should be asking me, surely.'

The interviewee also expressed his frustration towards the time the Ombudsman takes to deal with complaints:

'Well, the ombudsman service is terrible anyway. It takes months and months before they even acknowledge you, and then, usually, they don't, you know, it then takes several months, again, after they've acknowledged and they're going to start processing the complaint. It used to be you had to wait eight weeks, I think they recently changed that, but it was always a lot longer than eight weeks. It can be six to nine months before you hear back. … So it's a long process, far too long, far too distant, those are the biggest problems with it.'

The interviewee also felt that the service did not take the time to engage with him properly: *"I think, a couple of times, they do call you and say, 'Oh, I've just been assigned to this', etc., but it's no more than that, like, an interaction, and then you don't get anything else after that."*

He had a clear idea of what would make engaging with the Ombudsman a better experience:

'I've got disabilities and stuff and I get fatigued, and, you know, going through reams and reams of documents, sometimes I think I'm telling them too much, sometimes I think I'm not telling them enough. If they could say to me, you know, they're the professionals, "Look, you've said this point, you're not explaining it, you've not given evidence for it", or whatever. If they came back to me and had a discussion about it, you know, I'd be much more clear about what my role is, what I'm supposed to do, rather than just an angry rant. Do you know what I mean? If you could see them, you know, if we could have a meeting like this a few times during the process, and they were saying, "Look,

we've got to this point, what are you actually saying there?" That would be a much easier process, you know? And it would feel more as though it was to do with me, you know?'

Step 8 (Outcome)

The interviewee was clear about the limitations associated with what a help-seeker can achieve by going to the Ombudsman for support with their housing issue:

> *'Well, it's not really about trust, it's about recognizing their limitations, you know? I mean, even when you win. … The landlord doesn't comply, you tell them that they haven't complied, and you don't hear anything, you know? The frustration is recognizing the ombudsman is distant, is not engaged, and has limited hours even when they are, you know? To me that's just not, you know, appropriate, they should be seriously condemning the ombudsman, forcing them to face action, overseeing that action, and giving serious compensation, and looking to see that they make the changes that they are telling them they should do.'*

Therefore, it appears from the interviewee's account that the Ombudsman does not check to confirm the outcome of the complainant's case; namely, if what they have recommended is being implemented. The ombuds usually have a relationship with the LA and anecdotally their recommendations are complied with. One way the ombuds can hold the LA to account is to publish their decisions on the website for the public to see them.

The interviewee also explained that a lot of residents are unwilling to complain to the ombudsman due to the practicalities involved in the process that are difficult to navigate:

> *'It's my thing, I tell everyone to complain about everything, you know, first to the landlord, and then if they don't get a response, tell them. And a lot of people are just so frustrated with the process, they don't. I mean, around the service charge strike, we've still got a couple running out of the 15 people, and I told each one of them, "Please make a complaint". But people kind of get despondent and, like, you know, "What's the point, we're not going to win". I think that's a practical thing, you know? If they were more engaged, and you could see them more clearly, what their role was, I think more people would be willing to, you know, go through the process. But when they're so distant, and it takes so long anyway, it's kind of like, you know.'*

However, the interviewee noted some positive changes that have happened since, including the removal of the compulsory MP filter in the process:

'I mean, they've just made these changes, where you can go to them straight away. Previously, you had to go to a dedicated person, like an MP or a councillor, to get them to deal with it before their deadlines, and they've just changed that. So that might be interesting, to see if more people then go forward, because, you know, eight weeks is quite a long time if you've got some emergency repair going on, do you know what I mean it? The landlord's just dealing with it and they're just not complying with their own complaints process, or whatever, waiting another eight weeks, you know, is just too long. So people might, you know, start going there quicker, hopefully.'

The ombuds process is complex and takes time. There has been a debate for a while about whether to remove the MP filter from the ombuds process as it poses an additional obstacle for the user to go through. The process for the Property Chamber is different.

Property Chamber

In the following we describe a participant's experience going to, and through, the Property Chamber to resolve their housing issues.

Step 1 (Awareness)
The respondent described the problem he was facing with fellow residents.

'Our issue is to do with the overall management of our estate. Because we're a mixed use estate, we can't access what many other leasehold estates can do, which is right to manage, or enfranchisement. We just don't have those options because we have more than 25% commercial foot space on the estate.'

Step 2 (Taking action)
This respondent jumped Step 2 and went straight to considering whether to take their complaint to the Housing Ombudsman or Property Chamber, after briefly consulting with the advice sector.

Other respondents jumped Steps 2, 3 and 4 and went straight to considering whether to take their complaint to the Housing Ombudsman or Property Chamber.

Step 3 (Advice sector referral, support and guidance)
He sought advice from Leasehold Knowledge Partnership – an advice organization that advises leaseholders on the perils and pitfalls of the leasehold system and on how to succeed in a system that is rigged against them by commercial interests.

Step 4 (Intermediate processes)
This particular respondent did not complain to the organization involved and instead began to consider where to take his complaint. Like this interviewee, respondents did not always complain to the organization involved and instead went directly to either the Housing Ombudsman or the Property Chamber.

Step 5 (Consideration)
This respondent decided to take their problem to the Property Chamber.

> 'The only thing we could do is go to the tribunal to apply to have a tribunal-appointed manager. It's called Section 24 of the 1987 Act where the tribunal says that it's just inconvenient to remove the landlord as the manager or person with management responsibility of the building and put in a court appointed manager. It's the leasehold equivalent of putting in an administrator when a company goes into financial difficulties and an administrator is appointed.'

At this point in the process, participants reported either taking their problem to the Property Chamber or the Housing Ombudsman.

Steps 6, 7 and 8 (Engage, Service and Outcome)
Going to the Property Chamber created more problems for this particular user.

> 'In terms of what stage we're at, unfortunately we're in a process that seems to be largely never ending. We're a large, mixed use estate, and we first went to the tribunal when we had our hearing, in 2016. We were successful in that hearing but it has kicked off a whole lot of opposition from the landlord and we've been in and out of the tribunal ever since. ... That kick started a whole lot of appeals and arguments over the nature of the management order. Which is the order under which the Section 24 manager is appointed. Then also, the order was only for so long. It doesn't go on indefinitely and so what has happened, that all got started before the pandemic.'

The situation became more challenging as a result of the pandemic:

> 'Being in and out of the tribunal ever since and then the pandemic coincided with what would've been the end of the manager's term. We've had to apply to extend the order. Now, that's one thing that has just then been caught up in the pandemic, everything is taking that much longer. The Property Chamber, I don't know about where else in the country, but in London it just closed its doors and for a while, I don't know if it was doing anything quite frankly, it was impossible to tell. You weren't having remote hearings, there was just this silence. Actually, because then we had another application that coincided with the pandemic. That was we're applying to challenge the buildings' insurance,

so by contrast to the Section 24, this in theory is quite a simple application. It's saying, "We are being charged too much for our buildings insurance". It's a well-known issue in the industry because landlords and brokers take a massive commission. So, what happens, for the buildings insurance and just to summarize it, the broker, the landlord and the insurer effectively split the premium between them. The insurer allows the broker to keep 50% of the premium that is then shared with the landlord. So, imagine if your premium, which is as ours is, is half a million pounds, £250,000 of that is actually for the insurance. The other £250,000, well it turns out £75,000 went to the broker and the rest all went to the landlord.'

Due to the pandemic the whole process took two–and–a–half years to resolve:

'It has taken us, partly because of the pandemic, it took us two-and-a-half years to actually get the disclosure and this was the big, big issue. This actually is another issue in terms of access to justice, is that our tribunal, we have the same judge dealing with all of our matters, was just not actually engaged in the disclosure. How could I put this? We asked for the commissions to be disclosed and we kept getting back nothing. Or, we'd either had a statement that said, "The landlord doesn't know anything about commissions". Now, you wouldn't expect a judge to say, "Yes, fine, okay. The landlord doesn't know anything about commissions". You'd expect a judge to see that and say, "Hmm". But our judge didn't, our judge said, "Disclosure has been made". We had to apply for a third-party disclosure order to get the broker to disclose it. Because we had no in-person hearings and most of this is being done therefore by paper. Virtually, it took an awful long time and when we eventually had an oral hearing in February this year. So, I first made the application in December 2020 for the disclosure which is just before lockdown kicks in. It wasn't until February 2022 that we had an oral hearing at the brokers' request. … We're still waiting for the decision and that was done online. That was just two months ago, slightly less than two months ago.'

This participant also reflected on the difficulties someone more vulnerable would face, should they have been in his position:

'Yes, it's just a complicated process and if I'm educated up to, well accountancy and an MA and what have you. I used to be director of finance at the NHS, management consultant at KPMG. So, I consider myself to be an intelligent and articulate-, but even I have, you know, grasping how it works and how the system is played. I've got six years' experience now, but when I just realize now that what a lemon, how green I was when I first went into the tribunal. Of course, that's another disadvantage because it shouldn't be that you're permanently in and out of the tribunal in order to negotiate it.

Massive, because as I said at the start, this isn't necessarily about justice, it's settling legal arguments and that person is just not going to be aware, or it's very difficult for them to become aware. ... The law is just phenomenally difficult to navigate and there's no getting away from that. Actually, no matter how you try and it's the same with leasehold. You can't provide a simple and foolproof guide because there's always something that pops up that ...'

In fact, he suggested that a lot of people would actually just drop off the process because it takes so long:

'Sorry, yes. That's a very, very good and important point. That's one of the things that landlords do is they just grind you down. It comes a war of attrition, ours for example, appeals everything. ... No, absolute war of attrition in terms of wearing you down. The number of times I get, "Gosh, I couldn't do what you do, I would've given up and all the rest of it". That's exactly it and I think that's another thing about the legal imbalance in terms of the resources. It's very easy to get worn down, spun around. One example I'd give, is on a big estate therefore, it's possible for a landlord to rip off everybody, just say by £200. Now, for the landlord, that can yield £100,000, £150,000. But for an individual, to go to the tribunal, to fight for £200? I know people do it, but I would not personally. And so they get away with it and I think that's another thing.

The reason being is the complexity of the process has stopped them from doing anything. So actually, they've seen my experience and they've said, "No. Can't do what you've done". So, they just haven't bothered at all. They've just been shut out. In terms of technology, by and large, the groups, the cohort I'm moving in are all very well teched up.'

He specified that the greatest 'access issue' relates to leasehold law and its complexity:

'The first problem is leasehold law is very complicated and that's one thing you learn going to the tribunal. From my perspective, I think that's the biggest access difficulty, because it's interesting that you headed this about justice. I think people go to court, or they think they're going to court to access justice, but I would say that's true of a criminal court, but it's not true of a civil court in terms of people think they're going to get justice, but actually you're settling a legal dispute. When you go to a criminal court, you know what's happening, it's guilty, not-guilty and you know if justice has been served or not. When you're going to something like the Property Tribunal, it's not a case of wrong or right, it's what the law says about the issue that you have raised. Most people going into a tribunal don't quite get that because you may feel that it's blatantly obvious that something, an injustice has been done. But if your lease

or the legislation says something slightly different, you discover you've actually got no recourse. That's the big complexity, that actually you may go to the tribunal thinking, "Well, this is quite straightforward isn't it? Have I been overcharged?" Actually, it's not straightforward because it depends what your lease says and what the law says.'

This participant had experienced both remote and in-person Property Chamber hearings, which allowed him to reflect on the benefits and challenges of remote hearings.

Benefits included better access to materials and the comfort of accessing these remotely at home:

'But it was held remotely and to be honest, again, from an access to justice, actually that favoured me. Because I've got a massive 27-inch screen, I've got my iPad and what have you. So, access to documents and seeing the people, whereas in the court room, or the tribunal room, I'd have a tiny little laptop, because all bundles are electronic now. You haven't got the big A4 files. I just have a laptop and that would be it. Actually, I would put as a benefit, if you're properly equipped, which I am, actually a remote hearing puts you possibly at an advantage. It's actually easier to have your papers around you, your computer and what have you and have access.'

Benefits also included comfort breaks, something that would not be feasible in person:

'Actually, so when you then get the down time, the little breaks. So, shall we have a comfort break now? You don't tend to get ten-minute comfort breaks when you're in court, but you can do that online. As soon as you have that online, and you have your comfort break, your team-, so people who have been watching can immediately dial in on FaceTime and say, "That was a good point". Whereas, if you're in a court room, you'll huddle out and then you've all got to huddle back in again and so it all takes a bit more time. But I would add that, we're probably people in a privileged position in that we've got broadband, we've got computers and all the rest of it. It's an absolute nightmare if you're not in that position.'

Furthermore, lots of people could join in on remote hearings, something they could not have done in person due to social-distancing rules: *"Yes, absolutely. We had in our hearing in September, from our side we had seven or eight people who merrily could log on and so they turned their camera and their microphones off and they just sat and watched, yes."*

However, remote hearings do not allow for appellants to pick up on the nuances of the courtroom:

'I think yes, apart from the nuances of picking up who is doing what in the courtroom. I think it's impossible to do that. I don't know what the judges have, whether they just have one screen or two screens in front of them. But to have your papers and the screen, you can't see that somebody is shaking their head or frowning. Just those little subtleties are missed in the court room. … When we had this hearing in February, again an online hearing to do with the disclosure issue. I'd done my bit, it was then over to the landlord's witness or the broker's representative. I'm just sat here watching and listening and the other judge who is female, interrupted the person to ask them a question. Our judge immediately said, "Miss Jezzard (ph 16.29), please, you've had your say-, blah, blah, blah. Please could you let so-and-so-", I'm just sat there going whatever and the other judge is like-, then suddenly he says. "Oh, oh. That wasn't Miss Jezzard". No, and that for me, that wouldn't have happened in the court room probably because obviously you would've known it was his colleague sat next to him questioning.'

Conclusion

We have thus far presented a map of the pathways to seek help for housing issues that has shown the ideal case of how advice and justice can be accessed. We have also explored how those people who go through the process experience it. However, we have not included the 'professional perspective' in this chapter. We will include reflections from advice sector professionals, judges and case handlers in subsequent chapters.

The interview data examined here has exemplified that it is not straightforward to know where to turn to for help – most people do not know how to access these pathways, which leaves the system (more) accessible for those who are savvy and can navigate it. It is important to look at the role of community support and the role of the advice sector, as well as the overlaps in jurisdictions of ombuds/tribunals because it is often unclear which service a help-seeker needs to access. However, it is a promising development that the two can work together to signpost/cross-refer to each other. The complex and siloed system leaves the help-seeker in a vulnerable position if they are not able to navigate it. A lot of people give up, but our interviewees commented on the invaluable support from specialist charities and intermediaries in their complaint journeys.

The next chapter discusses the pathways through the AJS in the SEND context.

Notes
[1] Brighton Leaseholders Association (2022). www.leaseadvice.org/
[2] 'Guinness Housing Association Limited (202121926)', Housing Ombudsman Service (27 June 2022). www.housing-ombudsman.org.uk/decisions/guinness-housing-associat ion-limited-202121926/

3   Housing Ombudsman Service, 'Understand the complaints process'. www.housing-ombudsman.org.uk/residents/understand-complaints-process/
4   This series of videos will help you make a complaint to your landlord. Housing Ombudsman Service, Dispute Resolution Process: '1. Reporting an issue to your landlord' https://youtu.be/ZwbvhkHD_zY; '2. Making a complaint to your landlord' https://youtu.be/guTXv-1L1uo; '3. Escalating your complaint' https://youtu.be/9YpVUzriz0g; '4. The final stage of your landlord's internal complaints' procedure' https://youtu.be/so3HPzkkzp8
5   Housing Ombudsman Service (see n 3).
6   This series of videos will explain how to make a complaint to the Housing Ombudsman if you remain dissatisfied after complaining to your landlord and a designated person: '5. Referring your complaint to the Housing Ombudsman Service' https://youtu.be/wozF lG6O37I; '6. How the Ombudsman can help to resolve your complaint' https://youtu.be/9kgj_oWUu18; '7. Summary of the process' https://youtu.be/JxogqlrtEyo; '8. What are early resolution and local resolution?' https://youtu.be/QM8zyVDFj00.
7   Housing Ombudsman Service, 'Make a complaint'. www.housing-ombudsman.org.uk/residents/make-a-complaint/
8   See Housing Ombudsman Service, 'Factsheet on early resolution'. www.housing-ombudsman.org.uk/useful-tools/fact-sheets/early-resolution/
9   See Housing Ombudsman Service, 'Factsheet on investigation'. www.housing-ombudsman.org.uk/useful-tools/fact-sheets/investigation/
10 www.lease-advice.org/advice-guide/alternative-dispute-resolution-2/
11 A lease is usually made between two parties: a landlord and a tenant. However, it is also common for there to be a third party to the lease, such as a management company.
12 A list of available forms can be found here: Leasehold Advisory Service, 'Forms for England'. www.lease-advice.org/downloadable-form/forms/?topic=administration-charges
13 Leasehold Advisory Service, 'Application to the First-tier Tribunal (Property Chamber)'. www.lease-advice.org/advice-guide/application-first-tier-tribunal-property-chamber/

References

Brown, M. and Booth, R. (2022) 'Death of two-year-old from mould in flat a "defining moment", says coroner', *The Guardian*, 15 November.

Department for Levelling Up, Housing and Communities (2022) 'Press release: new deal for private renters published today', 16 June. www.gov.uk/government/news/new-deal-for-private-renters-published-today

Genn, H. (1999) *Pathways to Justice: What People Do and Think about Going to Law*, London: Bloomsbury.

Housing Ombudsman (2023) 'One year on follow up report: spotlight on damp and mould – it's not lifestyle'. www.housing-ombudsman.org.uk/wp-content/uploads/2023/02/Damp-and-mould-follow-up-report-final-2.2.23.pdf

McKeever, G., Simpson, M. and Fitzpatrick, C. (2018) 'Destitution and paths to justice: final report', York/London: Joseph Rowntree Foundation/Legal Education Foundation. https://research.thelegaleducationfoundation.org/wp-content/uploads/2018/06/Destitution-Report-Final-Full-.pdf

5

Pathways Through the AJS: SEND

Introduction

Using a similar approach to that of Chapter 4 (on housing), this chapter will explore the pathways to redress available to people through mapping (the ideal case) help-seeker journeys for people seeking help with special educational needs and disabilities (SEND) to understand how access points have been compromised and which pathways to justice are difficult to negotiate or blocked (Genn 1999; McKeever et al 2018). The Local Government and Social Care Ombudsman (LGSCO), the Parliamentary and Health Service Ombudsman (PHSO) and the SEND Tribunal provide redress for SEND problems. This cohort of administrative justice system institutions do not typically interact well with one another, and so we identify here the impact that digitalization could have on the procedures for these institutions to cooperate more effectively. There then follows a presentation of the ideal case help-seeker journey for those in need of support for SEND problems, using our specially developed journey map.[1] Finally, we will draw on interviews conducted with advice sector professionals, judges, case handlers and users to show how the help-seeker journey unfolds in reality.

Pathways to resolve grievances: ombuds and tribunals

The LGSCO and the First-Tier (Special Educational Needs and Disability) Tribunal provide redress for special educational needs problems. Although both provide redress for SEND issues, each deals with different aspects of a challenge. The LGSCO deals predominately with issues about the (in)actions of local authorities (LAs) in delivering the education, health and care (EHC) process. This includes areas such as complaints about the delay in assessing a child; about issuing the plan; and about failure to carry out reviews.

If there is a route to appeal to the tribunal, such as a decision not to assess a child; or on the content of the EHC plan, then the ombuds is not allowed to investigate these issues. They also do not have the powers to look at what happens inside an educational setting relating to special educational needs provision. Unlike the LGSCO, the SEND Tribunal deals only with *decisions* LAs make about children and young people with SEND and with schools that discriminate against a young person with disabilities.

Common SEND appeals to the tribunal include but are not limited to: refusal by the LA to undertake an assessment; refusal by the LA for SEND provision through an EHC plan; refusal to offer a placement in a specialist school.

Mediation

Another redress mechanism for SEND is mediation (Doyle 2019). Although this is an additional step in the journey when appealing to the tribunal, appeals *can* be resolved at this stage of the process. In the statutory framework of the Children and Families Act 2014, elements of mediation were made compulsory whereby parents and young people need to consider mediation before appealing to the tribunal and get a certificate to show that they had considered mediation. Unsurprisingly, because of the change to legislation, mediation increased exponentially from 75 cases in 2014 to 5,100 in 2021 (Marsons 2022).[2]

However, Cullen et al (2017) found a 14 percentage point reduction in the likelihood of an appeal being registered after mediation: 22% of appeals that have been mediated proceed on to the tribunal; and, of those who did mediate, 36% went on to an appeal. A parent/young person is likely to proceed to an appeal when there are multiple matters in dispute. For example, while issues around provision may be resolved, some questions, such as placement in a school, might not be. Dissatisfaction has also been expressed by parents relating to the engagement of LAs at mediation sessions.

In an interview with a SEND mediator, it became apparent that LAs are so stretched that they may come to a mediation underprepared. The LA representative needs to have decision-making authority: *"In mediation, we have to have a conversation with both parties before, help them work out what their position statement is, share that between the parties before the mediation, so they're coming in advance knowing the issues to be discussed."*

During the pandemic there was a step change in mediation as the process went online: *"We had lots of disputes with local authorities about whether they could use Zoom or not. Zoom is just a much better platform for mediation for lots of reasons, but even that just became a source of conflict."*

Interviewees recalled that the real challenge was to get the LA representatives to communicate with the mediators before the mediation. In many cases they appeared in the online meeting without having done the necessary preparation. Despite this, most mediations are still online today with only a few face-to-face mediations taking place. The LAs have come to appreciate the ease and speed of conducting mediations online. Further, the online format allows the decision-makers in the SEND department, representatives from schools and other professionals (for example, therapists) to be present (which can sometimes cause problems in an in-person session, due to busy schedules) and to take decisions. If the mediation is not successful, then the route to the ombuds or tribunal is still open.

LGSCO versus the SEND Tribunal

SEND issues relating to failures to follow policies and procedures, flaws in decision-making, poor administrative justice, and not considering an individual's specific circumstances might be dealt with by either the LGSCO or the SEND Tribunal, but the LGSCO cannot consider matters where the parent or carer has a right of appeal to the SEND Tribunal. Additionally, the LGSCO does not have the ability to investigate academies or schools' decisions or actions. The LGSCO and SEND Tribunal deal with different parts of a SEND challenge. It is difficult for a user to understand which types of complaint/appeal are handled by which institution. In fact, a user may need to apply to both the LGSCO and SEND Tribunal to deal with different parts of a complaint/appeal, making the process even more lengthy and difficult to navigate. Examples can be seen in the publication of LGSCO's SEND investigations which set out the parts they are unable to deal with. For instance, in the investigation against Kent County Council (August 2022):

> Miss D exercised her right to appeal to the SEND Tribunal about the content and the educational setting listed in F's EHC plan. The courts have said where the period out of education coincides with an appeal about an EHC plan and there is a link between them, the period from the date on which the appeal right arises until the appeal is heard is outside the Ombudsman's jurisdiction. Therefore, the Council's actions regarding F's education and specialist provision from 4 June 2021 to 4 January 2022 are outside of the Ombudsman's jurisdiction.[3]

This example demonstrates the complexity of the SEND appeals system and a need for better interaction between the two institutions to ensure a more streamlined process for their users. It shows how complex the distinction between the LGSCO's and the Tribunal's jurisdiction is. No matter how much LGSCO interact with the Tribunal, there will remain significant gaps in redress due to the way their respective jurisdictions work and the law that applies to both bodies.

In sum, the LGSCO deals predominately with the council[4] when it is alleged the council has failed to appropriately implement a child's EHC plan. The SEND Tribunal also deals with councils (and schools),[5] but unlike the LGSCO the SEND Tribunal handles only the *decisions* local councils make about children and young people with SEND and with schools that discriminate against a *disabled* young person specifically.

Complaints/appeal process: LGSCO, PHSO and the SEND Tribunal

This section presents the steps a person has to take when dealing with the LGSCO, the PHSO and the SEND Tribunal. Note that it would also be worth the complainant getting advice/checking legal aid eligibility via the Civil Legal Advice gateway.[6] The processes presented here are as described on each institution's website.

Making a complaint to LGSCO[7]

The steps through the process include: complaining to the organization involved; taking a look at the things LGSCO can and cannot look at; the timing of the complaint; registering a complaint; how the LGSCO will handle the complaint; and what the outcome might be.

1. *Complain to the organization involved*[8]
   The first thing the complainant should do is complain to the responsible council to give the council a chance to sort out the problem. The complainant should go through all stages of the organization's complaints process.
2. *Check the things LGSCO can and cannot look at*[9]
   The complainant's complaint must be about something LGSCO can investigate. LGSCO look at complaints about most council services, all types of adult social care services (even when the care is paid for privately) and some other organizations providing local services.
3. *Check it is the right time to complain to LGSCO*[10]
   If the complainant has completed the organization's complaints process but the complainant is not happy with its response, they can put in a complaint to LGSCO. If the complainant has complained to the organization but has not had a response within a reasonable time (up to 12 weeks) the complainant can also put in a complaint to LGSCO.
4. *Register a complaint*[11]
   The complainant must register an account on the LGSCO website and complete the online complaint form. LGSCO have procedures to provide assistance if there is a need for reasonable adjustments and the like.
5. *How LGSCO will look at your complaint*[12]
   LGSCO will take a look at the complaint and advise on the next steps. Then they will assess whether they can and should investigate. If they investigate, they may ask the complainant and the organization for more

information. LGSCO will ask the complainant if they need extra help to use their service, and do their best to communicate with the complainant in the way they have requested. This is predominantly a desk-based exercise rather than face-to-face contact.

6. *What the outcome will be*[13]

LGSCO will make an evidence-based decision on the complaint. If they decide the complainant has suffered because of the organization's faults, they will recommend how the organization should put things right for the complainant and potentially other people in the same situation. LGSCO publish their decisions, but do not use real names or reveal the identity of those involved. They do this to be transparent, increase accountability for what has happened, and to share the learning from complaints to help others improve.

Applying to the SEND Tribunal involves a different set of steps. We outline these in the next part, after briefly turning to the PHSO.

Making a complaint to the PHSO

The PHSO can be contacted when a parent or carer is unhappy with the delivery of the health provision in the EHC plan. The local health service that is being complained about needs to be contacted first.[14]

1. *Complain to the organization you're not happy with first*

If you're not happy with the service you have received from the NHS or a UK government department, let the organization know, so that it has a chance to put things right. If your complaint is about the NHS, you might want to start by contacting its Patient Advice and Liaison Service (PALS)[15] or complaints department. If your complaint is about a UK government department or public organization you will need an MP to refer the complaint to the PHSO.[16] You can complete the PHSO's complaint form[17] and ask an MP or their office to sign it. MPs will consider all complaints, no matter how big or small – from problems with a benefit or tax office to concerns you have about the Driver and Vehicle Licensing Agency or an immigration issue. If you have made a formal complaint to the organization and still do not feel the matter has been resolved, then get in touch with the PHSO as soon as you can.[18]

2. *Complain to the PHSO*[19]

The PHSO have a 'complaint checker tool'[20] that they will encourage you to fill in. It only takes a few minutes but may save you time by making sure that they are the right organization to look into your complaint. It will also ensure that your complaint is ready for the PHSO to look at. To submit a complaint you will need to download the online complaint form.[21] You will then have to provide information about the nature of the complaint, when it happened, how it affected you, what you would like done to put things right, if you are planning to take or have taken legal action, and if you are complaining for someone else.[22]

3. *Assessing your complaint*[23]

First, the PHSO will do some initial checks to make sure they can deal with your complaint. This includes checking that they can look into the organization and the issue you are complaining about and that you have been through the organization's own complaints process already. Second, the PHSO will then assess individual complaints to decide if they should investigate. For example, they might see that an organization has made mistakes, but it has already done what it could to resolve the complaint. Third, if it looks like there is a problem that still needs looking into, the PHSO can investigate. They then collect the facts to establish what has happened, weigh up the evidence and make a final decision on the complaint.

4. *PHSO can get things put right*

If the PHSO finds that an organization has got things wrong, then they can ask the organization to take action to put things right. This can include giving you an explanation or an apology. The PHSO can also ask the organization to take action to try to stop the same mistakes happening again. If they decide there were no failures, or that there were but the organization has done the right thing to resolve the complaint, the PHSO will explain why.

Applying to the SEND Tribunal[24]

The process of bringing a case to the SEND Tribunal is divided into the following steps: mediation; making an appeal; starting an appeal; before the hearing; the hearing; and after the hearing.

1. *Mediation*

The applicant should think about mediation before they appeal. This is when someone independently tries to help the applicant and the council resolve their problem. Even if the applicant does not use mediation, in most cases the applicant will need a certificate from a mediation service before they appeal. The applicant must ask for this within two months of the date on their letter from the council. The applicant has 30 days after the date on the mediation letter to appeal to the SEND Tribunal.

2. *Making an appeal*

The applicant must appeal to the SEND Tribunal within two months of the date on the letter telling them the council's final decision. The SEND Tribunal is free, and the applicant can claim money to pay for their travel to a hearing. The applicant might need to collect evidence to prove why they think the school or council is wrong. Some people can get money to help pay a solicitor for help with this. The Law Society or Citizens Advice can tell the applicant more about this.

3. *Starting an appeal (the 'appeal form')*

From the list of court and tribunal forms on the website, the applicant will need to download the correct form for appeals against a decision not to carry out an EHC assessment or appeals against any other local council decision. When the applicant appeals, they must tell the SEND Tribunal the date of the letter from the council and which of the decisions the applicant disagrees with. The applicant cannot just say they disagree with the decision. The applicant does not have to provide a lot of detail, but it is important that they explain the grounds of their appeal. This means explaining why the applicant thinks the decision is wrong and what they want the SEND Tribunal to do.[25]

The applicant must then post the appeal form or submit it electronically as an attachment to an email. If anything is missing, the tribunal will send the form back to the applicant and might not be able to look at the applicant's appeal. If this happens, they will tell the applicant what else the applicant needs to send them. The applicant must send the additional material within ten working days. The applicant *can* ask for more time, but, if the applicant sends it back late and does not tell the tribunal why, the applicant's appeal will end.

4. *Before the hearing*[26]

The tribunal will check the applicant's form to make sure their appeal meets the rules. The tribunal will do this in ten working days. Then they will write to say they have received the applicant's form and give the applicant an appeal number to use if they need talk to them about their case. The tribunal will tell the applicant the fortnight when the hearing will take place, and the tribunal will tell the applicant when the applicant needs to send the council and tribunal information for the hearing. The tribunal will send a copy of the applicant's appeal to the council and ask the council to reply within 30 days. If the council agrees with the applicant's appeal – for example, to change the EHC plan – the applicant can stop the appeal. If the council agrees to do anything else the applicant has asked for, then the appeal ends. The council has a set time to do what they say they will.

5. *The hearing*

About ten days before the date, the SEND Tribunal will send confirmation of the date and time of the applicant's hearing and tell the applicant where it will be. The SEND Tribunal try to make sure the hearing is less than two hours away from the applicant's home. A judge will lead the tribunal, and there will be one or two other people who know about children with SEND, health and social care matters. The judge will explain what will happen at the hearing. The applicant can agree to a hearing where they do not attend. But if it would help the tribunal to hear what the applicant thinks and ask the applicant questions, then the tribunal will not agree to a hearing on the papers.[27]

6. *After the hearing*

The tribunal will write to tell the applicant their decision and send a copy to the council. The applicant should get this within ten working days after the hearing has finished. The council must do what the SEND Tribunal says within a set time. There are different times for different decisions. If the council does not start when they should, the applicant can ask the Secretary of State for Education to tell the High Court to make them do it. When the tribunal writes to tell the applicant their decision, they will also say how to appeal. If the applicant is not happy with the decision, they must write back to the tribunal to tell them this within 28 days of the decision. The applicant must tell them what they think was wrong and why they want a new decision. If the applicant does it later than this, they must explain why. A judge can then decide whether the appeal can go ahead although it is late.

In the next section we present the ideal case help–seeker journey for complainants seeking help with SEND problems using our journey map. It is followed by real-life case studies of the different SEND pathways.

The ideal case help–seeker journey: a fictional case study

Bringing it all together, we now outline the ideal case help–seeker journey for those in need of support for SEND problems, using our journey map. We provide a case study of the SEND pathways from Step 1 (awareness) to Step 8 (outcome), and we use the same entirely fictional case study used in Chapter 4 to help illustrate some of the problems that people have and what they are required to do in order to access the system to seek advice, help and resolution.

Marta's SEND journey

Problem: Marta's son, Thomas, has special educational needs (SEN). Thomas has been struggling to learn at school and has also been experiencing bullying by other children. Marta wants to find help for her son, so she asks his school for assistance. She speaks to the Special Educational Needs Coordinator but is still not happy with the school's response, so she looks elsewhere for support.

Marta travels through the SEND pathways:

Step 1 (Awareness)
Marta wants to know what she can do about this. She asks her friends and searches online. She wants help to find out how she can get extra support for Thomas.

Step 2 (Taking action)
Marta has found specialist advice organizations online that might be able to help with her problem. She contacts them to find out how they can help.

Step 3 (Advice sector referral, support and guidance)
One organization tells her that Thomas might be entitled to an EHC plan. An EHC plan is a document that sets out a child's special educational, health and social care needs. The organization gives Marta advice and explains how she can apply for an assessment by the LA.

Step 4 (Intermediate processes)
Marta asks her LA to carry out an EHC needs assessment so that Thomas might receive help at school. The LA writes to Marta to say it has decided not to carry out the assessment. This means Thomas will not receive the extra help at school that Marta thinks he needs.

Marta reaches out again to the advice organization. She is told she has several options to seek help. The specialist organization will assist Marta to choose the pathway that is best for her specific problem.

We now discuss two options: Marta can *either* take her problem to the SEND Tribunal *or* go to the LGSCO. First, we will accompany Marta to the SEND Tribunal.

Step 5 (Consideration)
Marta goes back to the advice organization for more help. They tell her that she could appeal to the SEND Tribunal or try mediation with the LA first. In the letter Marta received from the LA, there are contact details for a local, independent mediation service. Marta contacts the mediation service to see if they can help. The mediation service sets up an appointment for Marta with the LA.

Unfortunately, the mediator is unsuccessful in helping Marta and the LA resolve the dispute. Marta is given a mediation certificate. If Marta now wishes to appeal to the tribunal she must do so within the deadline – two months from the date of the LA's letter or one month from the date of the mediation certificate – whichever is later.

Step 6 (Engage)
Marta goes to the HM Courts & Tribunals website and downloads the appeal form. She fills in the form and posts it to the tribunal, enclosing the LA's decision letter and her mediation certificate. She also has the option to email the form, but Marta feels more confident completing the form by hand.

Step 7 (Service)
The SEND Tribunal receives Marta's papers and checks they are correct and in time. They write to Marta and the LA with a timetable for her case and a hearing date. The letter explains that Marta can choose to have a hearing or let the tribunal make a decision on written evidence only. Marta decides she would like to have a hearing.

Marta attends the hearing. There is a two-person panel made up of a judge and one other person who knows about children with SEN. A representative from the LA also attends. With the help of the advice organization, Marta has already explained in her tribunal hearing application why she thinks Thomas needs an EHC plan.

Step 8 (Outcome)
The SEND Tribunal emails Marta and the LA once they have made their decision. The SEND Tribunal has agreed with Marta and orders the LA to carry out a needs assessment on Thomas. Once the assessment has been completed, the LA decides Thomas needs an EHC plan which includes extra support for him at school.

We will now discuss the path to the LGSCO which Marta can take with a slightly different problem relating to the EHC plan.

Step 5 (Consideration)
Thomas now has an EHC plan, but Marta does not think his school is providing the support set out in the plan. Marta asks for help from the advice organization she spoke to before. She is told to make a complaint to the LA about Thomas' support. The complaint is not resolved, so Marta is advised to take her complaint to the LGSCO.

Step 6 (Engage)
Marta makes her complaint online at the LGSCO's website.

Step 7 (Service)
The LGSCO investigates the issue and asks Marta and the LA to provide information. The investigator considers the evidence received and gives Marta and the LA the chance to comment on their initial views about whether the LA has been at fault and the impact this has had on Thomas.

Step 8 (Outcome)

After a few months, the LGSCO makes a decision. It has decided that the LA had not made sure that the school was following Thomas's plan, and so it recommends that the LA works with the school to make sure the plan is followed in the future. The LGSCO also asks the LA to make a payment to Marta and Thomas to recognize the provision he has missed out on. In this example, Marta is successful. Thomas receives the extra support he needs and is much happier. He is much better supported, able to achieve more and able to fit in with his peers, allowing him to thrive at school.

Note that at Step 5 Marta has several different options to seek help. The advice sector will assist Marta to choose the pathway that is best for her specific problem. We have discussed two pathways: she can take her problem to the SEND Tribunal *or* to the LGSCO. One of the differences between these two processes is that the outcome of a tribunal is legally binding, whereas that of the Ombudsman is a recommendation that the authority complained about can choose to comply with (anecdotally there is usually compliance). We will explore the differences in duration, satisfaction and compliance with the processes in the next section (drawing on interview data).

The help–seeker journey in reality: reflections from users

As in Chapter 4, our representation of the help-seeker journey is linear. However, it is never as straightforward in reality. The help-seeker may navigate the process differently, missing out or repeating stages, and often multiple things happen alongside each other. The help-seeker might abandon the journey, or get stuck along the way. The help-seeker might engage with the process actively or passively, and the help-seeker's circumstances can affect their decision-making.

The data we draw upon for this chapter are the SEND Tribunal and LGSCO interview data and survey responses[28] of the PHSO and SEND Tribunal provided by users. We include two exemplary case studies from interviewees: one for the SEND Tribunal and one for the PHSO. For the sake of illustrating that the help-seeker journey differs across cases – as already noted, help-seekers may navigate the process differently, missing out or repeating stages – we draw on different user experiences throughout to illustrate that the journey differs case by case.

SEND Tribunal

In the following, we share a mother's journey to, and through, the SEND Tribunal during the pandemic, to illustrate the help-seeker journey.

Step 1 (Awareness)
An information leaflet that was circulated at the interviewee's daughter's school set off a sequence of actions that helped her reach the tribunal for support.

> *'Well, I stumbled upon the services during the latter of the pandemic, an email was sent round to schools advertising the OTP [occupational therapy practice] systems for those who weren't coping in the pandemic. School didn't share this with me, or parents, until later on. I think it came out in April, we got it in October, and I said, "Well, obviously my daughter has personally struggled through the pandemic, and this would have been more helpful when it was first published, really". So, once I'd made contact with the lady, they were able to help me, and advise me where I could go.'*

The interviewee reached out for the help that was offered to all children in the school to support them through the pandemic. This is a sign of her ability to seek advice and ask for help. In this way, the mother learns more about her and her daughter's rights and can start the process to enforce them. She is legally enabled and savvy, which includes knowing where to go for help.

Our survey data and interviews with other users revealed that the majority of respondents spent a lot of time trying to work out their own problem before they went to the ombuds or the tribunal.

Step 2 (Taking action)
The interviewee went on to explain how she accessed the justice system:

> *'As a result of that ... we were in the process of doing EHCP [Education and Health Care Plan] for my daughter, a parental one because the school didn't agree that she needed any SEN care, but we had the support of Kaleidoscope, which is the outreach services from the hospital in the community services, and also had a booking at XXX hospital for her. We also had intervention from Drum Beats, which is an autistic society. I also tapped into the Autism Network. So, I tapped in to everybody because she was struggling really bad, and when your child has got suicidal indications, you need support, and the school wasn't giving me that support at the time ... they didn't understand her autism.'*

It becomes evident that the interviewee has existing contacts with hospitals and charities that can support her and her daughter's specific needs. The way she describes how she 'tapped into' networks shows a confidence in reaching out and exploring different pathways to help her get the support her daughter required.

Step 3 (Advice sector referral, support and guidance)
The interviewee went to different places for help, including OTP systems for those who weren't coping in the pandemic:

> 'It [the information] went to schools, all schools in Lewisham borough. I think it was in conjunction with the local authority to help children through the pandemic. … It was accessible to all children. It was every single child that was struggling. It was a global thing that should have been sent out to all children, all parents I should say.'

She also approached the Special Educational Needs and Disabilities Information Advice and Support Services (SENDIASS):

> 'I went to SENDIASS, I had only found SENDIASS through the educational psychologist, I think she put me on to SENDIASS. I never knew all these services existed. When you've got a child with special needs, you don't know these services are there to support you. This is year 6 we're talking about beforehand. Who else helped me? As I said, the Educate Autism Society, we've had a lot of TAC [team around the child] meetings. XXX hospital has been a very good advocate, because she has been under the hospital since birth.'

She also noted that Facebook was particularly useful: *"Can I just say for the last one, as well, Facebook also. Tapping into a lot of Facebook groups."* In sum, she is confident in sharing her story and reaching out – she uses social media and shows digital capabilities.

Step 4 (Intermediate processes)
The interviewee did not go to mediation: *"We've been to the tribunal because when you first do an EHCP, a parental one, the majority of the time it gets refused. Then they want you to go to mediation, which we didn't go to because reading from Facebook posts, and speaking to other people now, it's very unlikely that you get anything from mediation."*

Step 5 (Consideration)
She was advised – by SENDIASS and by peers on Facebook – to take her complaint to the SEND Tribunal. Like this mother, other participants also reported being advised to either take their problem to the SEND Tribunal or the Ombudsman. Our survey data showed that lots of people heard about the Ombudsman/tribunal through an advice agency and some from the LA.

Steps 6–8 (Engage, Service, Outcome)
The interviewee described the process to us. Critically, the pandemic created the online-only option:

> 'This was another nightmare because everything is done online. You have to upload all your documents, your documents aren't there, you put them in order. You turn into a professional advocate. And it's hard because SENDIASS work part-time. You have to get to meetings every two weeks, or week, you also have the emotion of family, you have a family to look after when you've got all this paperwork, all this legislation you don't really understand.'

This shows that a person who is digitally savvy can still experience the process as cumbersome and carries a large responsibility for getting it all right.

The interviewee commented on SENDIASS, the organization that helped her to go through the process:

> 'They did, they did help me a lot, it was a lot of nights of three o'clock in the morning, four o'clock in the morning getting that paperwork together, speaking to the hospitals, continuing my day-to-day life with an autistic child, going to hospital appointments, looking after my eldest daughter.'

The interviewee went on to say that she spent a lot of time trying to sort out the problem before contacting the tribunal:

> 'Altogether it has taken me a year, probably about a year. A very long time because you have to wait, then you have to submit, then you have to wait, then you wait for them to have the results. You can't just phone up for the results the next day. It's just a waiting game.'

When asked about her trust in the tribunal, she responded:

> 'You know what? ... I had a lot of trust in the tribunal. Even though you've heard a lot of horror stories, when somebody is separate from the local authorities, well, the school, if anybody was to look at XXX case, they would see that this child is suffering. It's kind of so where we put in our paper, we had to voice on paper to say a holistic approach needs to be addressed and because we said to be holistic, that's why the tribunal listened. You know? It would be interesting to see how many cases the tribunal refuses, because the threshold, as I said, is so low. ... Yes, I mean, their report was very good. It was a very good report, but then we had reports from Drumbeat, you know, we're dealing with a kid who has suicidal indications. If a tribunal had ignored that, then I would be very concerned about the system.'

She can understand her child's rights and knows how to navigate the system and whom to ask for help. She was clear about where in the process the system should have worked better:

'I think that in a school, educational psychologists, councils that deal with SEN, need to be more accountable for the care of our children because parents shouldn't have to fight for primary and then they get to secondary school and secondary school thinks, "What is going on? Are the parents alive? We never heard anything from the primary school. They never said such things". There has to be accountability for how children are raised in our society. They are our future. In a sense that, "Did the pandemic help?" Maybe the pandemic helped because I would never have got that leaflet. I would never have known about an educational psychologist. So, if it wasn't for that then I wouldn't know what I know now.'

It is important to note that each participant's story and experience of going to the SEND Tribunal that was shared with us was unique. For example, another interviewee shared her three-year long struggle to get her daughter – who is autistic and has attention-deficit/hyperactivity disorder and dyslexia – the support that she needs.

'[W]e won a Tribunal case against our local authority about her education. After three years in crisis and barely surviving in a mainstream school, we finally have a place at a specialist autism school that caters to kids of her profile. While this is a massively positive development, I'm left feeling hollowed and emotionally and physically exhausted. Why did it have to go this far and why is there not more support for families with special needs children? Why did we have to battle for her needs to be recognized and met every step of the way?'

This mother described her experience as a 'battle against the system'.

'Battling to get an Education and Health Care Plan, battling to get one-to-one support in school, battling to have her school start delayed by a year as she was half the cognitive age of her peers, battling to get any kind of therapeutic support to help us get through the days. We were at breaking point and left with nowhere else to turn. The only way forward was to go to the Tribunal.'

She explained that this 'battle' included privately commissioned assessments from a psychiatrist, educational psychologist and occupational therapist, 483 pages of evidence, sleepless nights and hours of preparation, as well as thousands of pounds in addition to the six-figure sum for years of home schooling and other support. She deemed the process almost impossible for someone less capable or educated to navigate, should someone more vulnerable have been in her position: *"I mean, it was, even for us, being fluent in English, being educated with professional backgrounds, it was bewildering, it was massively stressful. I cannot imagine someone who doesn't have a university degree, might not be fluent in English, to be able to manoeuvre the system."*

PHSO

In this section, we share the same mother's route to the PHSO to illustrate a different help-seeker journey. This complaint involved her second daughter and was consequently a separate case to the aforementioned. We draw on other user experiences throughout to illustrate that the journey differs case by case.

Step 1 (Awareness)

She became aware of the problem when her daughter started refusing to attend school.

> 'When [her child] was in school … throughout her school she had issues with teachers not understanding her autism. It came to the punch where [her child] refused to go to school, and didn't go to school for about a year and a half. We complained about a teaching assistant because the language that they used for [her child] was very condescending, these were [her child]'s words. Very derogative to her, whereas with other children they would just let it go over their heads, but those who are more understandable about the English language, you know you can't say certain things, the persona, the way you hold your face, she would take that onboard and take it personally. So, we put a complaint into school about this. [Her child] wrote a complaint. The school dismissed the complaint, but we were also fearful for [her child]'s safety in school because it's her word against his word. We wanted to know the truth about this. A child doesn't go home and say these things. So, we put a complaint in.'

Steps 2–5 did not occur hierarchically. This mother is an example of when the help-seeker may navigate the process differently, in this case multiple things happen alongside each other.

Step 2 (Taking action)

The interviewee complained to her daughter's school first. She explained that when you have a complaint against the school, that is the process. You complain to the school governors first. If they say no, you then have to go externally to either the Information Commissioner's Office (ICO) or on to the PHSO.

> 'You've got to do a SAR [suspicious activity report] if you want the paperwork. Then you've got to go through to the ICO. Then from the ICO, if you're not happy with the ICO, then you've got to go to the MP. Then you've got to go to the Parliamentary Health Service. Which is a lot. I mean, I don't mind you guys actually looking at [her child]'s case and reference if you're allowed to. Because, you know, for me it's about learning from this process. I've actually

set up an advocate group on the basis that I don't want any other parent to suffer the way we've suffered as a family or a child. So I have a meeting with an advocate group for SEND kids.'

Step 3 (Advice sector referral, support and guidance)
The interviewee only learnt about this process from experience: *"But yet again, just reading papers, tapping into Facebook, speaking to people. But, you know, I don't know anyone who's gone this route."*

Step 4 (Intermediate processes)
She described the intermediate processes relating to the ICO.

> 'Then we went to the ICO because the school wasn't giving us the notes from the teachers. The school wasn't playing a partnership in this. So, we pursued it with the ICO. The ICO said there was a case, and the school should be more helpful. The ICO was also under a lot of pressure with paperwork, this, that and the other. I said, "You've got a child that's got suicidal indications here, we need help fast". So, they looked at it again a second time, and they said, "No, we're not going to pursue it any further". I was a little bit confused about that. One minute you said there's a case, the next minute there's not a case. I know there's workload pressures. Is this why you've closed the case? Now we've taken it to the parliamentary and health service ombuds.'

Step 5 (Consideration)
It is apparent that the ICO has their own internal complaint process. You can complain directly to them or you can take your complaint elsewhere and go on to the PHSO instead, which the interviewee did, having failed to resolve her issue with the ICO. Like this mother, other participants reported being advised to either take their problem to the SEND Tribunal or to the PHSO.

Steps 6–8 (Engage, Service, Outcome)
But prior to making a complaint to the Ombudsman the interviewee had to go through an MP. She expressed frustration about the compulsory MP step in the process:

> 'No, after going to the ICO two times, we went back to the parliamentary-, no, we went to the MP first with this form, because we fill this form in, and then they send it … it's a step in the process. If you're sending things, why can't you send things to the Parliamentary Ombudsman? I just don't understand why the MP gets involved in this step.'

The interviewee commented on the hospital and charities that helped her to go through the process to reach the PHSO:

'And you know, if this is a girl with suicidal indications then, you know, if she didn't have a strong family she wouldn't be here today. And this is the implication of the education system and paperwork. Luckily, we do have the hospital and we do have charities that have helped me through this.'

However, the interviewee explained that the case has yet to come to an end:

'Well, they said they were going to ring me back. I've been waiting three weeks and nobody's got back to me. Because I said to them that the ICO had told me verbally that there's nothing more they can do. So it's my word against the ICO's now. But then it's, kind of, I've done all this stuff for nothing probably. And schools are going to get away with that, you know.'

Interestingly, the interviewee's complaint could have also been dealt with by the LGSCO that usually deals with LA-type issues. However, the interviewee had not heard of the LGSCO. This interviewee's case illustrates that people do not even know which ombuds to contact because there are so many ombudsmen which makes it difficult to select the most appropriate one when they want to make a complaint.

When asked about her trust in the Ombudsman, she responded:

'The ombudsman- it's kind of hard because it's kind of like they've phoned me up and said that I've not followed the protocol and I had to challenge it and say that I think I did because obviously, I made that call to the ICO and they told me about you. So, if I hadn't made that call then I wouldn't know about yourselves. So, there has to be some merit in what I'm saying and also this is a case that as I'm saying it could be a failed case. It could be on BBC News now, you know? If it was the other way, if something did happen, we'd be on BBC News. Do we have to go there before we get answers? Before people are corrected on protocols? Because that was my next thing. To go to maybe a newspaper or to a news presenter and say, "Look-". It's not working, but you know, it's tiring. And it's impacted my health as well.'

Conclusion

All our interviewees stated the importance of support from specialist charities and intermediaries in their complaint journey. Some have developed practical toolkits to help parents navigate the system.[29]

Thus far we have presented a map of the pathways to seeking help for SEND issues that has shown the ideal case of how advice and justice could be accessed. We have also explored how those people who actually go through the process experience it in reality. As the interview data has exemplified, the process is not straightforward; in fact, most people do not know how to access these pathways, which means that the system is (more) accessible for those who are savvy and can navigate it.

Overall, across Chapter 4 (housing) and Chapter 5 (SEND), most interviewees told us about delays to the process – even the duration of the individual steps to resolve their problem (or even to get heard) are very long – and there is a lack of guidance as to what to expect to happen and when, which leaves most users very upset and exhausted with the process. There is also an added element of emotional stress that the process puts upon those who undergo it.

Furthermore, it is important to take into consideration the role of community support and the role of the advice sector and intermediaries, as well as overlaps in ombuds/tribunals because it is often unclear which service a help-seeker needs to access and how the two systems can work together to signpost/cross-refer to each other. The complex and siloed system leaves the help-seeker in a vulnerable position if they are not able to navigate it.

Part III of the book explores, through our empirical data, what help-seeker journeys look like and how we might learn about (non) access to justice by applying our theoretical lenses of procedural justice and digital legal consciousness.

Notes

1 See 'Introduction' to Chapter 3 for an outline of the map, its design and its focus.
2 See also 'Reporting year 2022: education, health and care plans'. https://explore-education-statistics.service.gov.uk/find-statistics/education-health-and-care-plans#dataBlock-bd6ff903-b581-4aeb-acf4-6184df022c59-tables
3 LGSCO, Kent County Council, 21 009 839. www.lgo.org.uk/decisions/education/special-educational-needs/21-009-839#point1
4 LGSCO do not deal with schools (other than occasionally making third-party enquiries). LGSCO do not have jurisdiction.
5 But in two distinct types of cases. The tribunal considers challenges to councils' decisions about the assessment and identification of SEN and provision. Separately, it considers claims of disability discrimination against individual schools and responsible bodies.
6 See GovUK, 'Civil Legal Advice (CLA)' for more info on this pathway. https://www.gov.uk/civil-legal-advice
7 LGSCO, 'Make a complaint'. www.lgo.org.uk/make-a-complaint
8 For tips on how to complain to the council or care provider before going to LGSCO, see LGSCO, 'Top tips for making a complaint'. www.lgo.org.uk/make-a-complaint/top-tips-for-making-a-complaint
9 For more information on the services LGSCO investigate, see LGSCO, 'What we can and cannot look at'. www.lgo.org.uk/make-a-complaint/what-we-can-and-cannot-look-at; and for guidance on the common types of complaint LGSCO receive, see LGSCO, 'Complaint fact sheets'. www.lgo.org.uk/make-a-complaint/fact-sheets
10 For answers to the common questions about making a complaint, see LGSCO, 'Frequently asked questions'. www.lgo.org.uk/make-a-complaint/faqs
11 For more information on how to register a new complaint and what to do if you cannot use the LGSCO online form, see LGSCO, 'How to register a complaint'. www.lgo.org.uk/make-a-complaint/how-to-register-a-complaint
12 For more information on how the complaints process works, see LGSCO, 'How we deal with your complaint'. www.lgo.org.uk/make-a-complaint/how-we-deal-with-your-complaint

[13] To find out what will happen to your complaint, see LGSCO, 'What the outcome will be'. www.lgo.org.uk/make-a-complaint/what-the-outcome-will-be

[14] See the PHSO website. www.ombudsman.org.uk/

[15] See NHS, 'What is PALS (Patient Advice and Liaison Service)?'. www.nhs.uk/nhs-servi ces/hospitals/what-is-pals-patient-advice-and-liaison-service/

[16] You can find MPs' contact details at parliament.uk.

[17] See PHSO, 'Complaint forms'. www.ombudsman.org.uk/making-complaint/compl ain-us-getting-started/complaint-forms

[18] There are some organizations the PHSO can't look into. See PHSO, 'If we can't help'. www.ombudsman.org.uk/making-complaint/if-we-cant-help; and see PHSO, 'How we can help you leaflet' which explains the role of the PHSO and the types of complaints they investigate. www.ombudsman.org.uk/publications/how-we-can-help-you-leaflet

[19] See PHSO, 'Complaint to us: getting started'. www.ombudsman.org.uk/making-compla int/complain-us-getting-started

[20] The 'complaint checker tool' can be found at PHSO, 'Making a complaint'. www.ombuds man.org.uk/making-complaint#complaint-checker

[21] See PHSO, 'Complaint forms' (see n 17).

[22] See PHSO, 'Filling in our complaint form'. www.ombudsman.org.uk/making-compla int/complain-us-getting-started/filling-our-complaint-form

[23] See PHSO, 'How we deal with complaints'. www.ombudsman.org.uk/making-compla int/how-we-deal-complaints#step-1

[24] See HM Courts & Tribunals Service, 'SEND37 – how to appeal an SEN decision'. https://assets.publishing.service.gov.uk/government/uploads/system/uploads/attachme nt_data/file/776348/send37-eng.pdf

[25] More information on how to fill in the sections of the appeal form can be found in HM Courts & Tribunals Service, 'If you are not happy with a decision about Special Educational Needs (SEN)' (EasyRead version of SEND 37, note 24). https://assets.pub lishing.service.gov.uk/government/uploads/system/uploads/attachment_data/file/831 386/send37-easyread-eng.pdf

[26] More information on the process before the hearing can be found in EasyRead version of SEND 37 (see n 25).

[27] There is a film on YouTube that shows what happens at a hearing of a SEND Tribunal; see Ministry of Justice, 'Hearings at the Special Educational Needs and Disability Tribunal' (11 June 2021). www.youtube.com/watch?v=ExNEpi-E4XI. See also HMCTS YouTube, Supporting online Justice, for a selection of films about online hearings. www.youtube. com/playlist?list=PLORVvk_w75Py6JClMOiiltyTjI2gyc81g

[28] We designed and distributed 11 surveys from June 2022 to November 2022. However, despite our efforts to mitigate the low response rate, the final dataset had significant levels of missing data, rendering it unsuitable for our planned analyses. We were only able to produce descriptive statistics of the user sample available (N=40), and to run limited analyses using the more robust PHSO case-handler sample. We use some of the open-ended text responses here.

[29] For an example of a toolkit for problem-solving put together by specialist charities, see 'Problem-solving toolkit'. www.lukeclements.co.uk/wp-content/uploads/2016/02/Tool kit-draft-2016-04.pdf

References

Cullen, M.A., Lindsay, G., Conlon, G., Cullen, S., Bakopoulou, I. and Totsika, V. (2017) 'Review of arrangements for disagreement resolution (SEND)', CEDAR/University of Warwick. https://warwick.ac.uk/fac/ soc/cedar/research/disagreementresolution

Doyle, M. (2019) 'A place at the table: young people's participation in SEND dispute resolution: final report', Colchester: University of Essex. https://aplaceatthetablesend.wordpress.com/final-report/

Genn, H. (1999) *Pathways to Justice: What People Do and Think about Going to Law*, London: Bloomsbury.

Marsons, L.G.T. (2022) 'SEND reforms: mandatory mediation in administrative justice', London: UK Administrative Justice Institute. https://ukaji.org/2022/07/07/send-reforms-mandatory-mediation-in-administrative-justice/

McKeever, G., Simpson, M. and Fitzpatrick, C. (2018) 'Destitution and paths to justice: final report', York/London: Joseph Rowntree Foundation/Legal Education Foundation. https://research.thelegaleducationfoundation.org/wp-content/uploads/2018/06/Destitution-Report-Final-Full-.pdf

PART III

Exploring Help-Seeker Journeys

6

Exploring the Role of Procedural Justice in Tribunals and Ombuds

Introduction

With this chapter, we now begin Part III of this book, in which we explore the help-seeker journey by presenting our empirical data, starting here with a discussion of the vignette experiment. In Chapters 7 and 8, we will examine our qualitative data. In Chapter 2, we reviewed the literature on legitimacy, trust and procedural justice, which demonstrated that most prior research has focused on the police. However, in that chapter we also drew out some reasons why procedural justice may be important in administrative justice. In this chapter, we consider the idea that experiencing procedural justice during interactions with tribunals and ombuds is important not only in shaping legitimacy, but also in influencing perceptions of process transparency, outcome fairness, satisfaction and willingness to engage with the system in the future. We also assess whether the findings are different for online and offline proceedings.

Given the dearth of UK-based research on how people use and think about tribunals and ombuds services, open empirical questions remain, particularly in the new era of technologically mediated online interactions. For this reason, we fielded an online experiment to explore some of the underlying issues that placed research participants in hypothetical online and offline conditions using textual vignettes. Participants were presented with a story depicting a tribunal or ombuds procedure involving 'Marta', which was either online or offline, and either embodied procedurally just or procedurally unjust principles of interpersonal fairness and decision-making. We based this experiment on our scenario that we introduced in Part II. We manipulated key aspects of the scenario (that is, levels of procedural justice) to assess the potential impact on people's perceptions of fairness and legitimacy, as well as their willingness to engage with the system in the future (see Appendix). We did this in six steps.

First, we tested whether the procedural justice manipulations 'worked'. We did this after each experimental manipulation by (a) fielding a scale of people's perceptions of the procedural fairness that Marta had experienced and (b) assessing whether these perceptions varied in predictable ways across the procedurally just and unjust experimental conditions.

Second, we examined whether procedural justice increases people's perceptions of the transparency of the process. How tribunals or ombuds work is likely to be unfamiliar to research participants, and this unfamiliarity may create a sense of uncertainty about *how things work*. We tested whether treating Marta with respect and dignity, allowing her a sense of voice, and making decisions in neutral ways might help to create a sense of clarity and transparency by generating a sense of trust in the process.

Third, we assessed the importance of fair interpersonal treatment and decision-making in people's perceptions of the legitimacy of the justice system. We examined whether people who read a version of the vignette where Marta was treated with respect and dignity – where she was given the opportunity to tell her side of the story and so forth – were more likely to believe the institution was moral, just and fair, and therefore that it was entitled to be obeyed and respected. Procedural justice has been shown (a) to be core to normative expectations about the appropriate application of power (that is, central to the legitimization process) and (b) to signal status and value within hierarchical group settings (people are motivated to view authority figures as legitimate when they feel connected to the group that those authority figures prototypically represent). We examined whether the effect of procedural justice on legitimacy generalizes to research participants reading about a *hypothetical other*. The research participants are not, themselves, being treated with respect and dignity; they are reading about another person being treated in procedurally fair ways. Does this help instil a sense of legitimacy among research participants?

Fourth, we examined the effect of manipulating procedural justice on outcome fairness perceptions. The outcome was always a 'good one'; Marta always received a favourable outcome. Despite this, we considered whether research participants believe that the outcome was fairer when Marta was treated according to principles of procedural justice, compared to when Marta was treated in procedurally unfair ways. It may be that procedural justice offers reassurance that an unfamiliar or subjectively unusual procedure is being conducted appropriately and is therefore likely to reach appropriate outcomes (van den Bos and Lind 2002). Put another way, if people believe that authority figures are 'doing the right thing' in terms of fair and respectful interpersonal treatment and fair and unbiased decision-making processes, they may infer that they are also 'doing the right thing' in terms of the actual decision reached.

Fifth, we considered whether the importance of procedural justice provided general satisfaction with the way Marta's case was dealt with and participants' willingness to engage in such a process themselves. In line with the ideas sketched out, we tested whether the procedurally just scenarios elicited higher levels of satisfaction and a greater willingness to engage in the process in future.

Finally, we considered whether the relationships posited varied across online and offline scenarios. While Creutzfeldt and Bradford (2016) identified a 'procedural justice effect' in their UK data, drawn from users of ombuds services, their findings related to a mixture of online, traditional telephone, written and, to an extent, face-to-face interactions. As a result, it is unclear how this translates into purely online interactions (especially in the tribunals). Procedural justice may be less important in the context of the financial and usually very outcome-focused nature of these interactions, where most service users are clear about what they want and what 'success' looks like to them. On the other hand, though, the unfamiliarity of the online context may mean that people are even more attuned to cues of procedural justice than in more traditional, and familiar, offline interactional settings.

Experimental study participants

Respondents were drawn from the 'Prolific' online panel of research participants, constituting a quota convenience sample that matches the UK's profile along the lines of age, gender and ethnicity/race. Respondents self-selected into the study and were paid for their time.

In terms of age, 11% were 18–24, 20% were 25–34, 20% were 35–44, 15% were 45–54, 20% were 55–64, 13% were 65–74, and 2% were 75 years or older. On self-identified gender, 51% were female, 48% were male, and 1% preferred not to say. In terms of ethnicity/race, 78% were White British, 8% were any other White background, 3% were Indian, 1% were Black African, 1% were Black Caribbean, 1% were Pakistani, 1% were Bangladeshi, 1% were Chinese, 1% were mixed White and Asian, 1% were 'other' Asian background, 1% were 'other' mixed background (aside from mixed White and Black African and mixed White and Asian), and the rest were either mixed White and Black African, some other ethnic group, or preferred not to say.

When it came to employment status, 51% were employed, 18% were retired, 11% were self-employed, 6% were 'homemaking', 5% were studying, 4% were out of work or looking for work, 3% were unable to work, and 2% were out of work and not currently looking. With respect to education, 15% had GCSEs (or equivalent) only, 15% had two or more A-levels (or equivalent), 35% had a Bachelor's degree (or equivalent), 19% had a Master's degree (or equivalent), and the rest had some other qualification. Some 71% had no longstanding illness, disability or infirmity, 24% had at least one, and 5% said 'maybe'.

Research participants were asked whether they had ever tried to access institutions to help solve their problems. A total of 79% said yes (21% said no). Research participants were also asked whether they had been involved in a civil or criminal court case in the past three years. A total of 96% said no (4% said yes).

Method

Sixty research participants were randomly assigned to read one of eight vignettes drawn from a 2 × 2 × 2 factorial design:

1. the first dimension was tribunal (the Property Chamber) or ombuds (the Housing Ombudsman);
2. the second dimension was online or offline; and
3. the third dimension was procedurally just or procedurally unjust treatment and decision-making.

It is important to reiterate that each participant read just one vignette – the analysis presented therefore compares how judgements of, for example, the procedural justice of authorities varied across respondents exposed to the different vignettes. It is also important to note that we included the scenario of an *offline ombuds* process. As mentioned in the introduction of this book, the ombuds process is mainly online. But for the purposes of the experiment, we needed to create an appropriate comparison for the *online* ombuds process; without the offline condition, it would be difficult to interpret what respondents made of the online process, since there would be no point of comparison. So we included a fictional offline process.

Outcome variables included perceptions of the fairness of the process and outcome of the hypothetical case, as well as broader perceptions of legitimacy of judges and courts (in the case of tribunals) and ombuds (in the case of ombuds). We used scales to measure most key outcomes. Using confirmatory factor analysis, we found that variation in responses to all the items used in a particular scale was driven by one underlying, unobserved factor or latent variable, which we can call 'procedural justice', 'legitimacy' and so on, and which we can model and extract for further analysis.

The scale of perceptions of procedural fairness had the following Likert (agree/disagree) indicators. To what extent did the people that Marta dealt with:

- 'Always did what they said they would';
- 'Understood Marta's problem';
- 'Treated Marta with respect and dignity';

- 'Were unhelpful';
- 'Were easy to get in touch with';
- 'Tried as hard to help Marta as they would anyone else';
- 'Were impartial';
- 'Gave Marta the opportunity to express their views before decisions were made';
- 'Listened to Marta before making decisions'; and
- 'Made decisions based on facts, not their personal biases or opinions'.

Process transparency was measured using a single Likert (agree/disagree) question: 'To what extent was the process transparent?'

The outcome fairness scale had the following Likert (agree/disagree) indicators:

- 'Marta got the outcome she deserved';
- 'The outcome of the case was explained clearly';
- 'Marta received an outcome similar to that obtained by others in their situation'; and
- 'The length of time it took to deal with the case felt appropriate'.

In the tribunal conditions, legitimacy was measured using the following Likert (agree/disagree) indicators:

- 'We have a moral duty to back the decisions made by Judges because Judges are legitimate authorities';
- 'We have a moral duty to support the decisions of Judges, even if we disagree with them';
- 'We have a moral duty to do what Judges tell us even if we don't understand or agree with the reasons';
- 'Judges act in ways that are consistent with our own ideas about what is right and wrong';
- 'We support how Judges usually act';
- 'Judges stand up for moral values that are important for people like us';
- 'Judges can be trusted to make the right decisions';
- 'I believe that Judges can be trusted to act in ways that take into account the interests of citizens'; and
- 'Judges have the right to make decisions that affect people's lives'.

In the ombuds conditions, legitimacy was measured by similar Likert (agree/disagree) indicators:

- 'We have a moral duty to back the decisions made by Ombuds because Ombuds are legitimate authorities';

- 'We have a moral duty to support the decisions of Ombuds, even if we disagree with them';
- 'We have a moral duty to do what Ombuds tell us even if we don't understand or agree with the reasons';
- 'Ombuds act in ways that are consistent with our own ideas about what is right and wrong';
- 'We support how Ombuds usually act';
- 'Ombuds stand up for moral values that are important for people like us';
- Ombuds can be trusted to make the right decisions';
- 'I believe that Ombuds can be trusted to act in ways that take into account the interests of citizens'; and
- 'Ombuds have the right to make decisions that affect people's lives'.

The experimental design means that when assessing the results of the study the statistical analysis is simple. Specifically, statistically significant group differences (tested via ANOVA or linear regression modelling) in mean levels of key outcome variables can be interpreted as the causal effects of manipulations. Due to the legal and other differences between the tribunals and ombuds services, we present our results separately for each (as, essentially, two 2 × 2 studies).

Before we summarize the findings, it is important to reiterate that participants were not recent users of tribunals or ombuds services. Because they were drawn from the general public, specifically by Prolific to help form their research panel, the experiment can be seen as providing a baseline. We consider how people *in general* think about the procedural fairness of online and offline procedures in a regulatory context that is unfamiliar to them (only 19 of 480 participants had been involved in a civil or criminal court case in the last three years, for example). Do they attend to questions of procedural justice in this context? What impact does procedural justice have on perceptions of process and outcome and on broader attitudes towards legitimacy and future potential engagement with the administrative justice system? Does it matter whether the procedure is online or offline? We also assume that these judgements and expectations would be relevant if participants were ever to use a tribunal or ombuds service.

Tribunals

We first found that the manipulation of procedural justice 'worked'. Specifically, there were large differences in procedural fairness perceptions comparing the procedurally fair and procedurally unfair conditions. People who read a vignette that described Marta experiencing respectful treatment, voice and unbiased and transparent decision-making thought that interpersonal treatment and decision-making were fairer, on average, than people who read a vignette that described Marta experiencing procedurally unjust interpersonal interactions and decision-making. This statistically significant comparison was the case whether the tribunal was offline or online, as can be seen in the visualization shown in Figure 6.1, which summarizes the raw data, the means (black dots) and the 95% confidence intervals around the means (black vertical lines above and below the black dot mean). Note that the comparisons between unfair and fair conditions are statistically significant (the 95% confidence intervals do not overlap) and the effect sizes are relatively (and unsurprisingly) strong. On a scale running from −2.4 to 1.2 (mean of 0), the means for the procedurally just conditions were 0.85 (online) and 0.85 (offline) and the means for the procedurally unjust conditions were −0.89 (online) and −0.83 (offline).

Second, and consistent with the findings, people in the procedurally just conditions tended to see the process as more transparent, compared to people in the procedurally unjust conditions (Figure 6.2). On a single-item scale running from 1 to 5 (with a mean of 3.2), the means for the procedurally just conditions were 4.3 (online) and 4.2 (offline) and the means for the procedurally unjust conditions were 2.3 (online) and 2.5 (offline). In some ways, this can be seen as an additional manipulation check, since transparency is part of procedural justice. But given that there are good reasons why transparency reduces one's sense of uncertainty, and given that the manipulation check in Figure 6.1 did not explicitly ask about transparency, it is important do this extra piece of analysis.

Third, we found a similar pattern for perceptions of outcome fairness (Figure 6.3). Even though the outcome was always the same 'good outcome', people who read a procedurally just vignette thought that the outcome was fairer, compared to people who read a procedurally unjust vignette. The findings were understandably not as strong as for fair process perceptions, since process fairness was manipulated and outcome was not. On a scale running from −3.6 to 1.4 (with a mean of 0), the means for the procedurally just conditions were 0.16 (online) and 0.37 (offline) and the means for the procedurally unjust conditions were −0.38 (online) and −0.16 (offline). Again, however, the comparisons between unfair and fair conditions are statistically significant (the 95% confidence intervals do not overlap).

Findings were less strong when considering perceptions of the legitimacy of judges and the tribunals. Perceptions of legitimacy were higher in procedurally just online tribunals than in procedurally unjust online tribunals. In offline tribunals, however, the comparison between procedurally just and procedurally unjust conditions was not statistically significant. The effect sizes can be seen in Figure 6.4. On a scale running from −3.4 to 1.8 (mean of 0), the means for the procedurally just conditions were 0.14 (online) and 0.32 (offline) and the means for the procedurally unjust conditions were −0.28 (online) and −0.19 (offline). The confidence intervals underline the point

Figure 6.1: Tribunals: raw data, means and 95% confidence intervals for procedural justice perceptions across the four experimental conditions

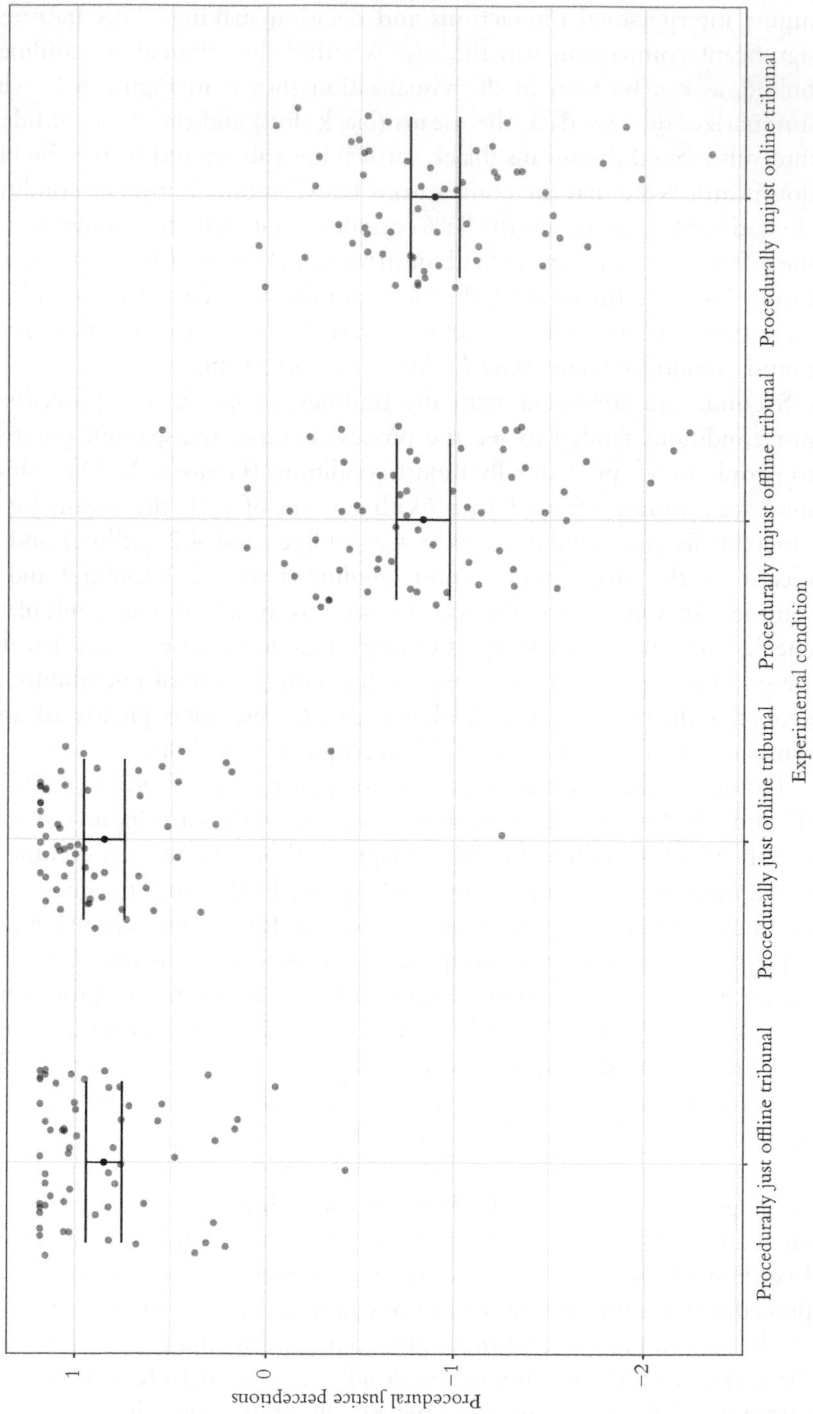

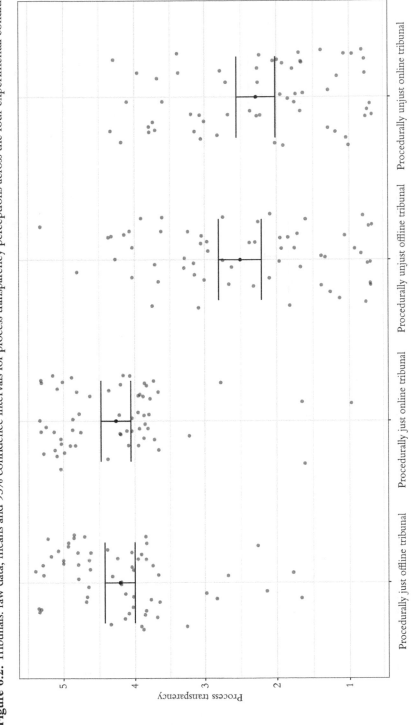

Figure 6.2: Tribunals: raw data, means and 95% confidence intervals for process transparency perceptions across the four experimental conditions

Figure 6.3: Tribunals: raw data, means and 95% confidence intervals for outcome fairness perceptions across the four experimental conditions

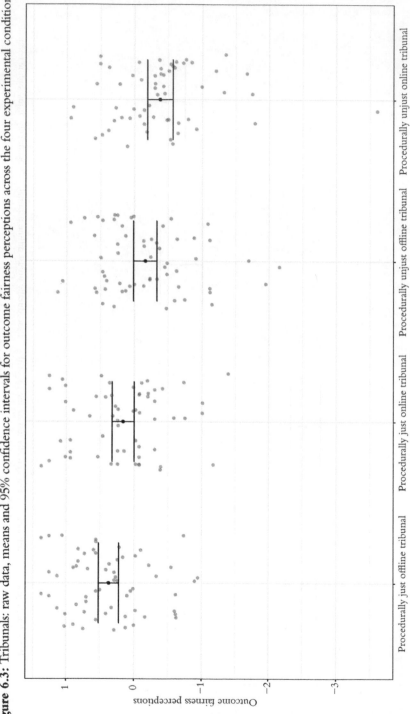

Figure 6.4: Raw data, means and 95% confidence intervals for court/judge legitimacy perceptions across the four experimental conditions

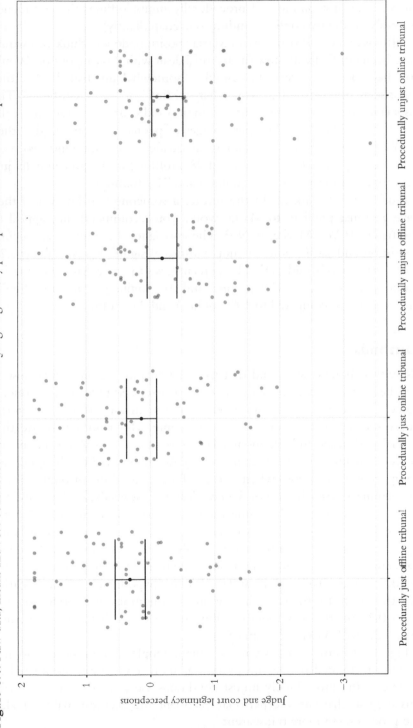

that the only comparison that was statistically significant was that between procedurally just offline and procedurally unjust offline (the confidence intervals for the two online conditions overlap slightly).

Fifth, we asked: 'If you were in Marta's position, do you think you would have been satisfied with how the tribunal dealt with this case or not? Would you have been ... "very unsatisfied", "somewhat unsatisfied", "neither unsatisfied nor satisfied", "somewhat satisfied" or "very satisfied"?' The association between experimental condition and satisfaction was statistically significant (Table 6.1). The percentages of people who agreed that they would be 'very satisfied' or 'somewhat satisfied' (so combining two rows in Table 6.1) were 98% (online) and 98% (offline) in the procedurally just conditions, compared to 58% (online) and 71% (offline).

Finally, we also asked: 'In the future, if someone you know says they have a similar problem to Marta, would you recommend they appeal to the tribunal? Yes, Maybe or No?' The association between experimental condition and satisfaction was not statistically significant (see Table 6.2). Note that nobody said 'no'. The percentages of people who said 'maybe' (rather than 'yes') were 15% (online) and 10% (offline) in the procedurally just conditions, compared to 19% (online) and 25% (offline).

Ombuds

As with tribunals, we found that the manipulation of procedural justice 'worked'. People who read a vignette that described respectful treatment, voice and unbiased and transparent decision-making thought that the process was fairer, on average, than people who read a vignette that described disrespectful treatment, a lack of voice given to Marta, and biased decision-making that lacked transparency. Again, this statistically significant comparison was the case whether the tribunal was offline or online, as can be seen in the visualization of the raw data, the means (black dots), and the 95% confidence intervals around the means shown in Figure 6.5. As with tribunals, the comparisons between unfair and fair conditions are statistically significant (the 95% confidence intervals do not overlap) and the effect sizes are, once again, relatively strong. The means for the procedurally just conditions were 0.87 (online) and 0.73 (in the hypothetical offline scenario) and the means for the procedurally unjust conditions were −0.86 (online) and −0.76 (in the hypothetical offline scenario), with the scale running from −2.2 to 1.3 (with a mean of 0).

Figure 6.6 shows that, as with tribunals, people in the procedurally just conditions tended to see the process as more transparent, compared to people in the procedurally unjust conditions. So just like with tribunals, this suggests that fair interpersonal treatment and decision-making makes the process seem more transparent.

Table 6.1: Satisfaction with tribunal proceedings across the four experimental conditions

Experimental condition	If you were in Marta's position, do you think you would have been satisfied with how the tribunal dealt with this case or not?					Total
	Very unsatisfied	Somewhat unsatisfied	Neither satisfied nor unsatisfied	Somewhat satisfied	Very satisfied	
Procedurally just online	0%	2%	0%	12%	87%	100%
Procedurally unjust online	5%	25%	12%	46%	12%	100%
Procedurally just offline	0 %	0%	2%	7%	92%	100%
Procedurally unjust offline	2%	23%	5%	54%	16%	100%
Total	4	30	11	71	125	241
	2%	12%	5%	30%	52%	100%

χ^2=144.517; df=12; Cramer's V=0.447; Fisher's p=0.000

Table 6.2: The possibility of recommending people appeal to the tribunal across the four experimental conditions

Experimental condition	In the future, if someone you know says they have a similar problem to Marta, would you recommend they appeal to the tribunal?		Total
	Yes	Maybe	
Procedurally just online	85%	15%	100%
Procedurally unjust online	81%	19%	100%
Procedurally just offline	90%	10%	100%
Procedurally unjust offline	75%	25%	100%
Total	199	41	240
	83%	17%	100%

χ^2=5.018; df=3; Cramer's V=0.145; p=0.171

Third, and compared to tribunals, we found a similar pattern for outcome fairness perceptions. The outcome fairness scale again had the following Likert (agree/disagree) items: 'Marta got the outcome she deserved', 'The outcome of the case was explained clearly', 'Marta received an outcome

141

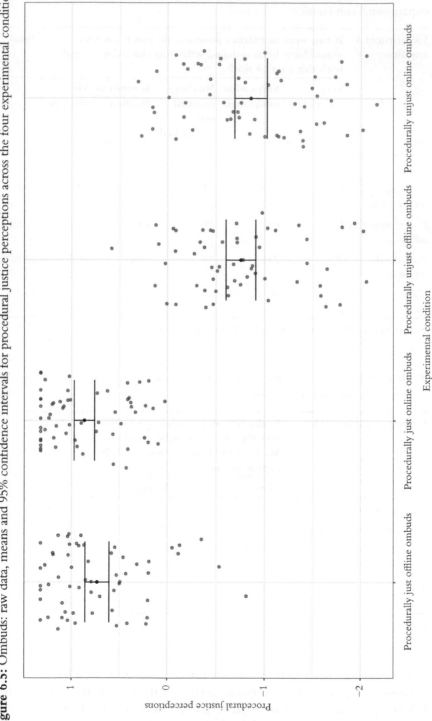

Figure 6.5: Ombuds: raw data, means and 95% confidence intervals for procedural justice perceptions across the four experimental conditions

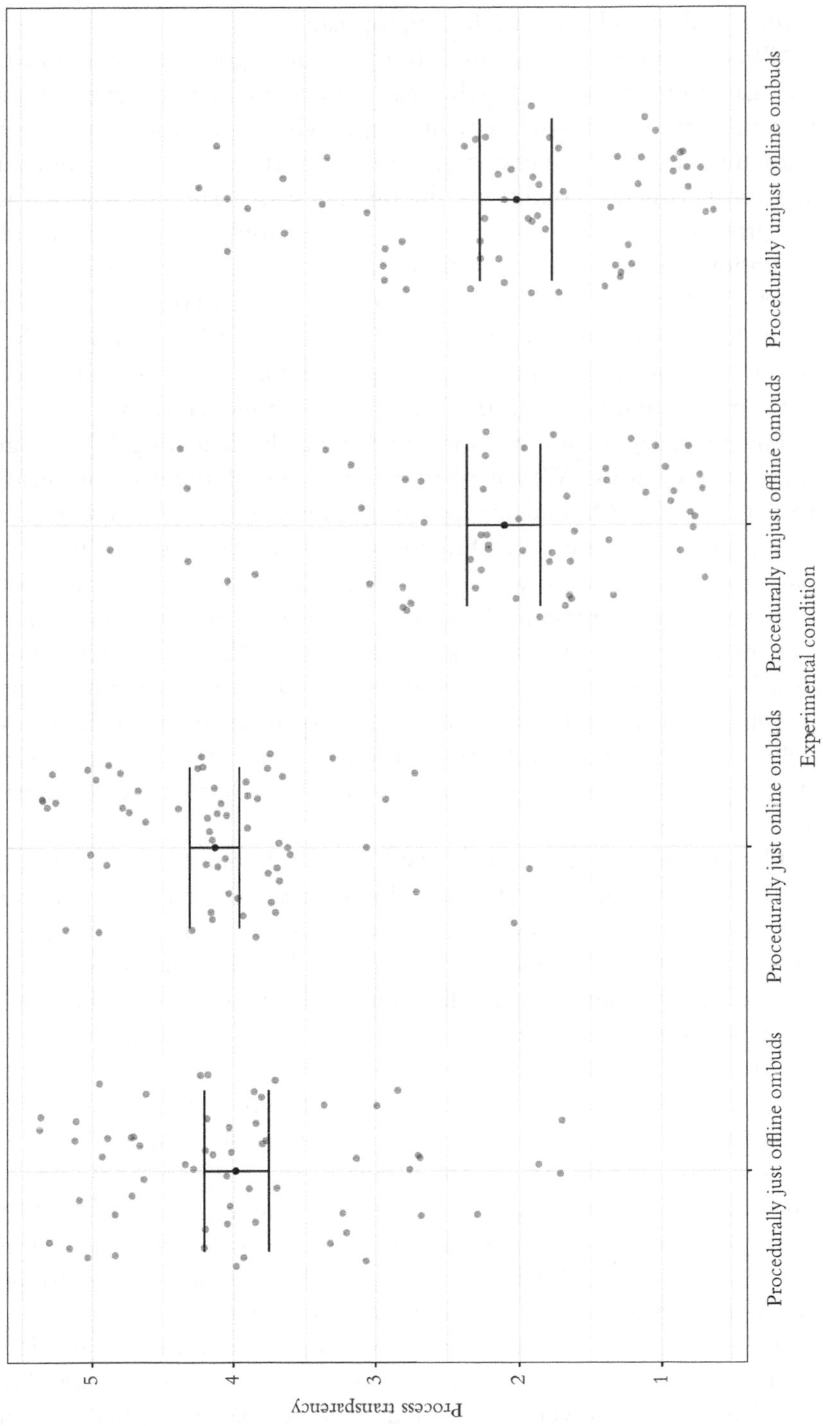

Figure 6.6: Ombuds: raw data, means and 95% confidence intervals for process transparency perceptions across the four experimental conditions

similar to that obtained by others in their situation', and 'The length of time it took to deal with the case felt appropriate'.

While the outcome was, again, always the same 'good outcome' for Marta, we again found that people who read a procedurally just vignette thought that the outcome was fairer than people who read a procedurally unjust vignette (Figure 6.7). Reflecting the fact that the outcome was not varied across experimental conditions, the findings were again not as strong as for fair process perceptions. Note that the comparisons between unfair and fair conditions are statistically significant (the 95% confidence intervals do not overlap). On a scale running from −2.0 to 1.8 (mean of 0), the means for the procedurally just conditions were 0.30 (online) and 0.33 (in the hypothetical offline scenario) and the means for the procedurally unjust conditions were − 0.31 (online) and −0.32 (in the hypothetical offline scenario).

This time, legitimacy was measured using the following Likert (agree/ disagree) indicators: 'We have a moral duty to back the decisions made by Ombuds because Ombuds are legitimate authorities', 'We have a moral duty to support the decisions of Ombuds, even if we disagree with them', 'We have a moral duty to do what Ombuds tell us even if we don't understand or agree with the reasons', 'Ombuds act in ways that are consistent with our own ideas about what is right and wrong', 'We support how Ombuds usually act', 'Ombuds stand up for moral values that are important for people like us', 'Ombuds can be trusted to make the right decisions', 'I believe that Ombuds can be trusted to act in ways that take into account the interests of citizens', and 'Ombuds have the right to make decisions that affect people's lives'.

The findings for legitimacy were, as for tribunals, less strong than for process and outcome fairness. In the reverse of tribunals, however, perceptions of legitimacy were stronger when comparing procedurally just and procedurally unjust *online* ombuds, but the comparison between procedurally just and procedurally unjust in the hypothetical offline ombuds was not statistically significant (with tribunals, it was *offline* that was significant not online). The effect sizes can be seen in Figure 6.8. On a scale running from −3.4 to 1.5 (mean of 0), the means for the procedurally just conditions were 0.28 (online) and 0.24 (hypothetical offline scenario) and the means for the procedurally unjust conditions were −0.39 (online) and −0.14 (hypothetical offline scenario). Again, the confidence intervals show that the only comparison that is statistically significant is that between procedurally just online and procedurally unjust online (the confidence intervals for the two hypothetically offline conditions overlap slightly).

We asked: 'If you were in Marta's position, do you think you would have been satisfied with how the ombuds dealt with this case or not? Would you have been … "very unsatisfied", "somewhat unsatisfied", "neither unsatisfied nor satisfied", "somewhat satisfied" or "very satisfied"?' As with

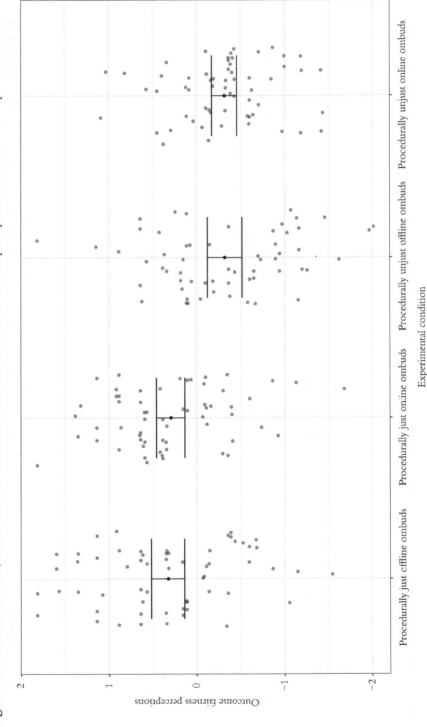

Figure 6.7: Ombuds: raw data, means and 95% confidence intervals for outcome fairness perceptions across the four experimental conditions

Figure 6.8: Raw data, means and 95% confidence intervals for legitimacy (ombuds) perceptions across the four experimental conditions

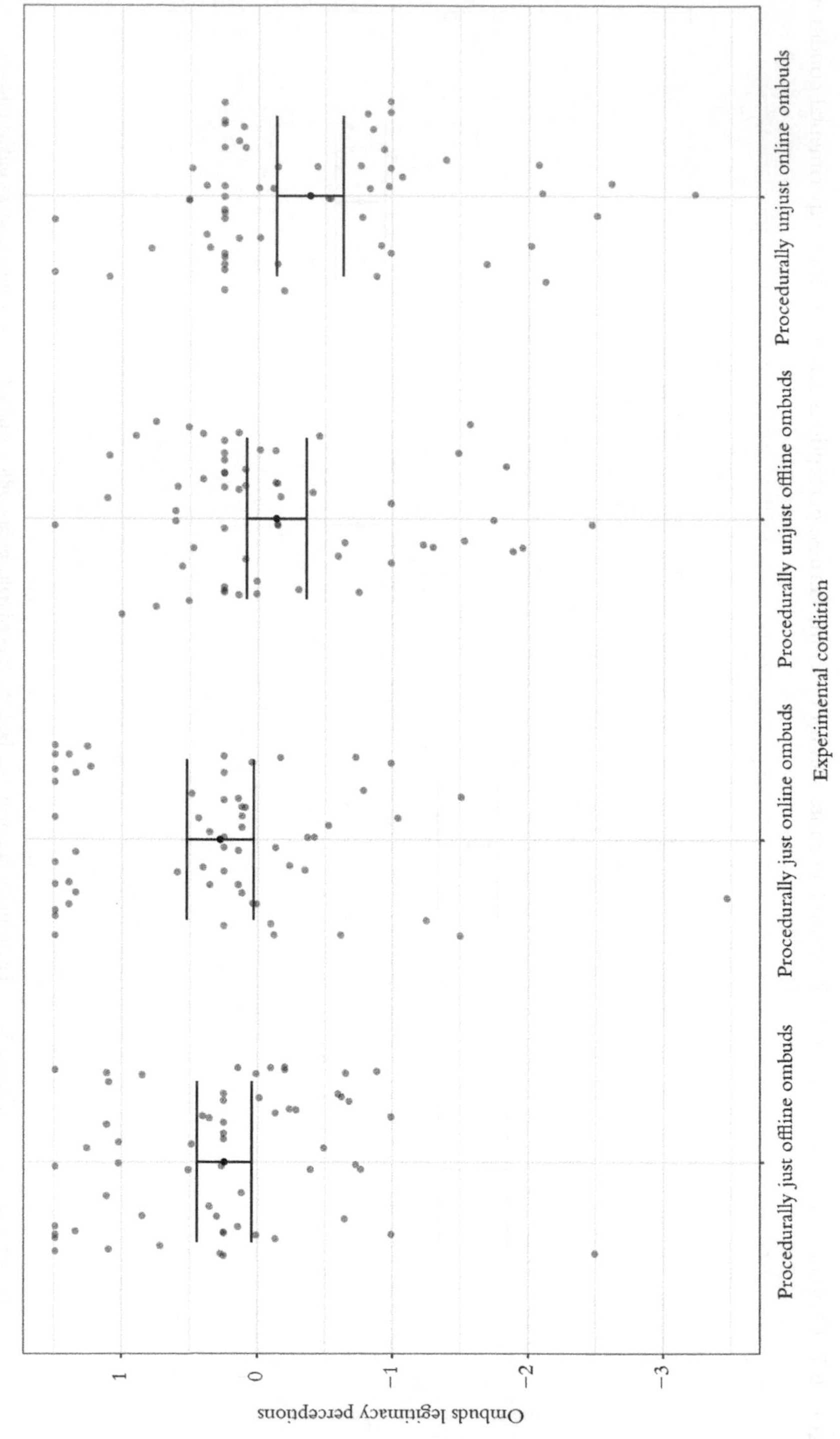

Table 6.3: Satisfaction with ombuds proceedings across the four experimental conditions

Experimental condition	If you were in Marta's position, do you think you would have been satisfied with how the ombuds dealt with this case or not?					Total
	Very unsatisfied	Somewhat unsatisfied	Neither satisfied nor unsatisfied	Somewhat satisfied	Very satisfied	
Procedurally just online	0%	2%	2%	15%	82%	100%
Procedurally unjust online	10%	32%	22%	32%	5%	100%
Procedurally just offline	0%	0%	5%	22%	73%	100%
Procedurally unjust offline	12%	29%	9%	41%	10%	100%
Total	13	37	22	65	102	239
	5%	16%	9%	27%	43%	100%

χ^2=139.416; df=12; Cramer's V=0.441; Fisher's p=0.000

tribunals, the association between experimental condition and satisfaction was statistically significant (Table 6.3). The percentages of people who agreed that they would be 'very satisfied' or 'somewhat satisfied' were 97% (online) and 95% (hypothetical offline scenario) in the procedurally just conditions, compared to 37% (online) and 51% (hypothetical offline scenario) in the procedurally unjust conditions.

We also asked: 'In the future, if someone you know says they have a similar problem to Marta, would you recommend they appeal to the ombuds? Yes, Maybe, or No?' The association between experimental condition and satisfaction was this time (compared to tribunals) statistically significant (see Table 6.4). Note that, once more, nobody said 'no'. Compared to tribunals, the findings were stronger: the percentages of people who said 'maybe' (rather than 'yes') were 10% (online) and 22% (hypothetical offline scenario) in the procedurally just conditions, compared to 42% (online) and 44% (hypothetical offline scenario) in the procedurally unjust conditions.

Discussion

Overall, four key findings emerged from our experiment. First, the procedural (in)justice manipulations worked for tribunals and ombuds, including, interestingly, making the overall process seem more transparent.

Table 6.4: The possibility of recommending people appeal to the ombuds across the four experimental conditions

Experimental condition	In the future, if someone you know says they have a similar problem to Marta, would you recommend they appeal to the tribunal?		Total
	Yes	Maybe	
Procedurally just online	90%	10%	100%
Procedurally unjust online	58%	42%	100%
Procedurally just offline	78%	22%	100%
Procedurally unjust offline	56%	43%	100%
Total	163	67	230
	71%	29%	100%

χ^2=21.993; df=3; Cramer's V=0.309; p=0.000

As we speculate in the final chapter, respectful interpersonal treatment and neutral decision-making may reduce the sense of uncertainty that could feel inherent in unfamiliar processes and situations. Second, despite the outcome being always a relatively 'good one', research participants said that the outcome was fairer when Marta was treated according to principles of procedural justice. We offer some thoughts on this in the concluding chapter. Third, perception of legitimacy was lower in procedurally unjust conditions compared to procedurally just conditions among offline tribunals (but not online tribunals) and online ombuds (but not hypothetical offline ombuds).

Fourth, levels of satisfaction 'if research participants were to be in the same position as Marta' were higher in procedurally just tribunals compared to procedurally unjust tribunals (again, despite the identical positive outcome) and even higher in procedurally just ombuds compared to procedurally unjust ombuds (despite the same positive outcome); and, when asked 'In the future, if someone you know says they have a similar problem to Marta, would you recommend they appeal?', people in procedurally just ombuds cases were more likely to say 'yes' than people in procedurally unjust ombuds cases, but there were no significant effects for tribunal cases. Finally, there was some indication that procedural (in)justice made a little bit more of a difference in the online compared to the offline scenarios.

Conclusion

Our experimental data suggest that in the current context, as in many other justice-related arenas, procedural justice is important in interactions between the users of ombuds services and tribunals and the authority figures involved.

It is therefore important that authority figures in these contexts – judges, case workers, and so on – are convinced of the importance of behaving in procedurally fair ways and given the tools to do so (that is, there may be a need for both generating awareness and delivering appropriate training).

Recall, Chapter 2 highlighted that applying the principles of procedural justice to ombuds and tribunals can lead to greater trust and confidence in these institutions. By ensuring that procedures are transparent, accessible and impartial, these institutions can demonstrate a commitment to treating all individuals fairly and justly, enhancing their legitimacy and credibility in the eyes of the public. This can be achieved, in part, through treating people with respect and dignity, giving them a sense of voice and participation in the process, and making fair and neutral decisions. Indeed, procedural justice may be particularly important here given (a) people's general unfamiliarity with the processes involved and (b) an increasingly online format. Under such conditions people may look especially for the reassurance that process fairness provides. At the very least, our data show that in online scenarios people pay just as much attention to cues of procedural justice as they do in face-to-face scenarios, and on balance it seems likely they pay *more* attention to procedural justice.

A further focus may be needed on the interplay between the interpersonal skills of service providers and whether they are operating in structures that allow procedurally just interactions to take place (tribunals and ombuds differ). Perhaps most obviously, as the move online gathers pace, there may be a temptation to use this to increase efficiency by reducing the time allocated to a particular hearing or procedure. Yet, our experimental vignettes demonstrated making a process fair, from the perspective of those involved or viewing it, may take time. For example, explaining what is going to happen in ways that services users can understand takes a little longer to do properly. 'Structural' elements of the process may therefore hinder (or enhance) the potential for procedurally just interactions. Those overseeing the structural context need to pay as much attention to the quality of the process, and how their actions affect this, as they do to the individual agents involved. Both need to take responsibility for change.

In the following two chapters we discuss our qualitative data and explore access to digital justice, marginalized groups and unmet legal needs.

References

Creutzfeldt, N. and Bradford, B. (2016) 'Dispute resolution outside of courts: procedural justice and decision acceptance among users of ombuds services in the UK', *Law & Society Review* 50(4): 985–1016.

van den Bos, K. and Lind, E.A. (2002) 'Uncertainty management by means of fairness judgments', in M.P. Zanna (ed), *Advances in Experimental Social Psychology*, vol 34, San Diego: Academic Press, 1–60.

7

Access to Digital Justice

Introduction

In Chapter 6, we discussed our quantitative experiments and concluded that procedural justice matters in the context of online interactions; this and the next chapter will discuss the question of access. How accessible is online justice? We ask this question in the context of the data discussed in Chapters 4 and 5, which explored people's interactions with the digital justice space. This chapter enquires into how those who administer justice, those who provide advice and those who use the online justice system experience it. In so doing, we consider how the use of technology in the justice system is shaped by, and may reshape, people's orientations and sensibilities towards law (Merry 1990; Cowan 2004; Hertogh 2004; Silbey 2005) and technology (Elvidge 2018). We use our data to examine, through people's (non)digital journeys, in what way consciousness of how we think and feel about the law relates to our capability of acting upon it. In Chapters 4 and 5, we have considered help-seekers' different levels of access, different levels of capabilities, and different levels of abilities in relation to (non)access. In Chapter 6, we discussed our quantitative experiment and showed the importance of procedural justice and legitimacy for online and offline encounters with the justice system.

In this chapter, we move on to the themes emerging from the interview data with judges, lawyers and case workers and contrast these with our user data through the lens of digital legal consciousness (digital and legal capabilities). The data we draw upon for this chapter are the SEND Tribunal and Property Chamber interviews and the survey responses[1] of the advice sector, case handlers, judges and users.[2] Based on the four basic types of digital legal consciousness (Chapter 1), we identified our user responses to be a combination of *Type 1* and *Type 2*, discussed in the next section. Examples of *Types 3* and *4* will be discussed in Chapter 8. As mentioned, there are no clear boundaries separating these types in real life; they merely serve here as a foundation to build upon. For example, a *Type 3* user could take part in a tribunal hearing, assisted by a representative or other supporter, but our dataset did not contain them. There are also fine-grained distinctions within and between these four types. For example, levels of digital and legal capabilities can change (even throughout the process of engaging with the legal system). We make sense of these complex interactions in the conclusion through bringing them into conversation with other lenses discussed in this book: namely, procedural justice and emotions.

Experiences of online hearings in SEND tribunals and property chambers

We found *Type 1* (digitally and legally savvy) and *Type 2* (digitally and legally savvy but preferring to keep legal processes in person and away from the online space) represented in our dataset. *Type 2* respondents had no choice during the pandemic: they had to access the justice system online. This makes them more critical of an online hearing but still perfectly capable of navigating it. All respondents recalled a range of emotional responses they had had during and because of their digital journeys. This suggests that there is an emotional dimension that influences how we think about the online justice system. These emotions form during, and because of, the experience of the online process and will play a role in shaping our digital legal consciousness. In other words, taking *Type 1* and *Type 2* as our starting point exposes a complex relationship between capabilities that equip people to navigate the online space and understand the legal dimension and the emotions that the process creates. This helps to draw out, and further probe, some imbalances of experiences of digital journeys.

In the following sections we present some of the professional interviewees' (judges, lawyers and case workers) responses in five themes that emerged from the data: advancements in technology and access to justice; ensuring inclusive justice; access to justice in the digital age; challenges and opportunities of technology in justice; and face-to-face hearings and trust in the legal system.

Advancements in technology and access to justice

Professional perspective

Most of our interviewees (judges, lawyers and case handlers) expressed a clear preference for online hearings and working from home. They are all equipped with the necessary technology, relevant training and a reliable internet connection to manage their caseloads and access online bundles from a home office. For example, a SEND judge commented on the impact online hearings had had on her life.

> 'From my point of view, my life has changed in the last two years. I've gone from not being able to do the school run and never seeing my child, to now

being able to do the school run, at least in the morning. With 300 miles a day I just couldn't do it. And I don't think that the courts have understood that, at the end of the day, our lives have changed and this has been our life for two years, and we don't want to go back to all that travel and never seeing our families again.'

Another SEND judge commented on the ease of holding different hearings in different places within a click of a button.

'In terms of the tribunals, the efficiency with which we can be deployed, I can sit with somebody from Newcastle and Brighton in the morning and people are off somewhere else and we've got a hearing in Manchester or whatever it happens to be. So no one's going to say "Well how long's it going to take them to get there?".'

When we asked a housing case worker about the benefits of working from home she said:

'I must admit, I like seeing people and hearing the evidence, and getting the opportunity. I do enjoy doing them [hearings] remotely, certainly for me it's a very convenient way of doing it, rather than having to travel up to London, but certainly when you do them in person, you get that extra, don't you, the body language, and you can see those sorts of things that you don't get on a remote setting definitely.'

These accounts highlight some of the benefits experienced by professionals; ease of working from a home office, no time wasted in the car, in traffic or in a hotel, which means less time away from family. Further, it was mentioned that remote hearings can be in different places on the same day, which is not easily possible for in-person hearings. Some professional respondents recalled that conducting hearings online saves time and money, plus they get paperwork done faster and have more quality time with their families.

A judge observed, from the point of view of a person bringing a case to the SEND Tribunal online:

'I get the feeling when people join and they're talking to you and they're sat there in the living room, they're sat on their sofa, they're in their environment, they're in a safe space for themselves. And I think it feels to me that they feel more able to talk more freely, and you can get a lot out of people with some really easy open questions. Whereas I just feel in the formalities of a face-to-face courtroom you might not get that same level of feedback. Again, I haven't got any empirical evidence for that, it's just the feeling I get.'

This observation, that people might feel more comfortable attending from their homes, is probably true for some but not all cases. We cannot forget that, despite hearings being remote and accessible from anywhere, they are about important decisions in people's lives. A Housing Law Practitioners Association member observed that: *"Judges felt like they were more able, or it seemed like judges were harsher in what they did, like granting possession when there wasn't a human being and maybe their family sitting in front of them and they weren't telling that person, 'You're now homeless'."*

As stated in the responses, the quality a computer screen can bring to a hearing is limited when it comes to body language or reading people's reactions to what has been said. However, on balance, factors such as who is needed at a hearing (panel members and schoolteachers for evidence for SEND cases, and experts for housing cases), and that they might find time in their schedule more easily if it is online, might have to be taken into account.

A SEND judge commented on this in relation to online and in-person hearings:

> *'If you're going to have an in-person hearing your panel needs to be together wherever they are. You have schools that have to send deputy headteachers, headteachers, that may spend a full day in a hearing room to simply answer one or two questions. It's a massive waste of their time, but they need to be there in case they are needed. If they could join remotely, that's a much better use of their time.'*

The same judge went on to say:

> *'I'm not averse to face-to-face sitting. I would do that where it's required. I have my own prejudices, in that respect. … It's the wider return to the office piece. I think there has to be a value in something that you're doing. You can't just do it for the sake of it, or because it was there before. For me, I feel I've worked fairly successfully from here [home].'*

The points made by the interviewees provide a snapshot of emerging imbalances. Our data suggest that judges feel at ease operating from their homes, as work can be more efficient and they have the equipment to conduct remote hearings. The decision to continue with remote tribunal hearings requires consideration of the economic benefits, such as increased efficiency, as well as the non-economic factors, such as the impact on individuals' personal lives and wellbeing, which are more difficult to quantify. While economic justifications are often viewed as more objective, the COVID-19 pandemic has brought to light the significance of affective considerations in making decisions about the future of remote tribunal hearings. Further, judges stated that people who bring a claim to the tribunal seem at ease in an online hearing from their own living room. There is another side to these observations, that of the tribunal users. The pandemic created a fertile ground for understanding better how online hearings and online communication with users worked and where the challenges lie. The experiences of tribunal users, with varying levels of legal and digital capabilities, did not always match the narratives about the benefits that the professionals expressed. The user perspectives are discussed in the next four themes, following the professional perspectives.

Ensuring inclusive justice

The professional perspective

The SEND judges we interviewed all commented on the benefits that remote hearings have had in enabling the service user and, in some cases, the young person whose case was being discussed to join in. They found that the overall experience was more relaxed than having to come in front of a court in a face-to-face setting. For example, a SEND judge stated that:

> 'The accessibility for our service users was fantastic. We're supposed to be an informal tribunal, so the parties were coming in, especially parents, who were feeling more relaxed, they didn't have to find childcare for their child if they're home educated. We could hear from the child or young person if they wanted to. It's amazing how many young people are coming, and children, coming to talk to us. And it's nice to put a face to the person who we're seeing. Because we wouldn't see them in court generally, we wouldn't want them to be in court, we don't want them to hear what can seem quite negative, for a whole day. … So it's nice that they can come in, come out, come say hello to us, give us their views, tell us something. Some children can't tell us anything, but it gives us a better viewpoint. And I think it's made me think how dangerous any judging is because you don't know that person who you're judging, you've got a snapshot.'

The fact that young people can join the online hearing, if they choose to, has benefits if this does not put additional stress on the parent or guardian arguing for better support for the child in an online hearing. As the spectrum of challenges a young person faces can be incredibly complex, it is not easy to make a claim that the online format is a good option for inclusion and access for all. However, we present two more accounts of the described benefits, as perceived by judges. First, a comment made by a judge about young children with autism:

> 'With a focus on the children who the decisions are about, those with autism, for example, find this easier as well, because it's like a video game for them.

*But also we can alter this in so many ways that we couldn't alter a court. ...
I think you can make it more accessible this way, than you ever could in courts.'*

Second, another SEND judge told us about one of their hearings:

*'I had the young person concerned come and join with his mother and his
grandmother, actually. As it happened it was a fairly uncontested matter in the
end, we ended up going to a consent order, but he was there, he was present.
We had a chat, as it were. But we certainly thanked him for turning up, and
it makes the experience a lot more pleasant to deal with the person that you're
helping decide their fate, effectively.'*

In sum, the judges in our dataset see online hearings as an advantage for
young persons, as they can choose to be part of the hearing; it also allows
the judge to put a face to the story they are deciding about. Our data clearly
show that, for the judges, seeing the young person provides for a far better
process and a more personal experience.

Tribunal user perspective

There is, however, also the side of the parent, carer, representative and the
young person themselves. Here, we present data that show a combination of
agreement and disagreement with the judges' statements. A SEND Tribunal
user, for example, stated that: *"[T]he video system was rubbish (May 2020) but
I heard they later changed it. It is hard focusing on a hearing whilst looking after a
child who was shielding so no one else could help. LA [local authority] people only
on phone, not video, which didn't seem fair."*

A Property Chamber user stated that:

*'It does depend on whether you've got the access, but you can just listen in.
You can just dial in on the phone and just listen, it's not so obvious then as
to who is speaking. ... I would add that we're probably people in a privileged
position in that we've usually got broadband, we've got computers and all the
rest of it. It's an absolute nightmare when you're not in that position.'*

Some of our respondents, though digitally and legally savvy, still experienced
challenges with the online process. The challenges were caused by their
surroundings and the reality of their lives; for example, caring responsibilities
and other demands on their time. This puts emotional stress onto an already
difficult situation and will undoubtedly influence how people perceive the
online process. The imbalance of visibility to an online hearing (to be heard
but not seen) clearly caused distress to the party bringing the claim. We
discuss this later.

An account that is slightly different, from a SEND Tribunal user, is that of an entirely paper–based hearing. She recalls:

> '[T]hey [the tribunal] were really good because sometimes they'd send documents out and I didn't really understand it because they wanted the council to prepare something, papers, which first when you read it you think, what have I missed? What have I done? I don't understand this. Every time I rang the tribunal, they were helpful. Obviously, the council knows the system very well. They save documents up until the last hour, not always inclusive of what they share with the tribunal with you.'

The account shows the reassurance and security a tribunal user felt with the option of calling a tribunal officer to help them navigate the system. They are not digitally savvy, not fully at ease with the process, and find reassurance in human guidance. They are fully aware of the power imbalance between themselves and the council whose staff are used to the system and process. SEND cases are about children who are currently not getting what they need from the system, and, no matter how digitally and legally savvy a parent might be, emotions and unexpected challenges of the online process can tarnish their experiences and play a role in forming their emerging digital legal consciousness.

Access to justice in the digital age

Professional perspective

Our professional interviewees provided us with some examples of the benefits for people to join a hearing remotely, from their homes. The data speak to the themes of vulnerability and access. The responses show some of the benefits of a remote hearing for those users that are confident with technology and have a good understanding of the legal process (*Types 1* and *2*). However, these people might have a vulnerability that prevents them from easily travelling to a tribunal or they might have other responsibilities that affect their ability to go to a face–to–face hearing. Here, we see clearly how other considerations, besides legal and digital capabilities, play a role in shaping people's attitudes towards the online justice system. For example, a law centre lawyer shared: *"Good is the fact that if you've got somebody who is disabled and can't travel it's good for them to be able to sit in their own homes."*

A Property Chamber case handler commented that:

> *"the benefits of face-to-face are that tech issues are less of a problem, but increased travel time, less accommodating for disabled, etc., video hearings exclude the digitally challenged but can save time, can more easily be recorded and reviewed in case of dispute, easier to accommodate some disabilities and caring responsibilities."*

Another advantage to attending a hearing from home, assuming that the person's internet connection is stable and that they are able to navigate the online hearing, is that it makes the experience easier for those people who have different variations of anxieties and fears about leaving their homes and interacting with people face-to-face. Similar to the judge's comments mentioned earlier, people noted that an online process saves time and money, as they do not have to embark on a potentially long journey to a courthouse. On the other hand, the fact that a hearing can be done from home also means that there is an abrupt end once the screen is shut down. There is no debrief, social interaction or ritual that would usually occur in a courtroom (Mulcahy and Rowden 2020).

A Tribunal Enforcement Services lead commented:

> 'I would always make sure to talk through the hearings I did with the tenant applicants afterwards about their experiences and I think on one level it's less scary if you get to be in your own room instead of being in a courtroom or tribunal hearing room. That's just a more anxiety-inducing prospect. They found it easier to actually not get flustered by the cross-examination process. This applied to landlords as well. So, the truth of the matter came out a little bit more, there was less opportunity for bullying behaviour I suppose by barristers, but there was a real practical element. A lot of the times the tenants that we represented, the reason why they were entitled to the action is at its root cause because they're very poor and they don't have any money. That's why they are in these housing situations. There's a lack of alternative and when you are maybe in that situation, it's not always possible to take a day off, arrange childcare, pay for transport costs to get into central London and it's really tricky because obviously for the people who are working in the tribunal, both as representatives, judges, that's not a lived reality they've got any idea of, but if you don't have the money to pay a babysitter ...'

Our data only touch upon some of the many different stories of people's experiences of an online hearing. There are also many reasons why an online hearing can be problematic. We might have to ask ourselves what is lost in an online hearing and for whom, as well as documenting the benefits of online hearings for the parties. And can the online space provide the same outcome and experience for users, at least for those that possess digital and legal capabilities?

Tribunal user perspective

The accounts of tribunal users with digital and legal capabilities to access and navigate the online hearing describe the benefits of being able to attend the hearing from home, but recall the time it took to get a judgment as a problem.

A SEND user reported that:

> 'Video hearings were much better for us. With our SEN children, not having to commute made it much easier to attend. There have been long delays in getting responses from the tribunal. We did a video hearing in April or May 2020 – it took more than four weeks and some chasing to get the judgment. I think the May 2021 judgment was quicker. Most recently, we submitted a consent order jointly with our LA in January 2022. It took until very late April 22 to get the consent order confirmed – meanwhile our daughter was left without the necessary support.'

Similarly, an Independent Provider of Special Education Advice (IPSEA) employee said:

> 'a positive which was a case where realistically a remote hearing was the only viable option for this family because the child, or might have been a young person, had complex needs including complex health needs that meant that arranging care for them if they were to physically have to travel to a tribunal venue would have been incredibly difficult. They had nursing support and there were risks that my colleague was saying, realistically it's really difficult to see how they could have taken part in the hearing if they all had to somehow get there. She couldn't envisage how it could have been done face-to-face in person.'

Another user reflected upon the sense of safety the online process provided for them, as compared to the possibility of having an in-person hearing. They articulated the tension between online and in-person hearings. A SEND Tribunal user shared with us that:

> 'a lot of the stuff has been done online, which if I had gone in person to a hearing, would I feel overwhelmed even more? Probably would because, obviously, you've got a qualified solicitor against the parents. The funding for that was another avenue that I explored about getting a solicitor involved. The fees are horrendous. Horrendous. People have spent £15,000, if not more, just fighting the systems. You know? It's not right. I'd rather spend that money on her care and her future. So, I think of doing things the way they do, but they need to be shorter.'

These examples illustrate the benefits of accessibility to an online hearing for people with different needs. Further, the pandemic taught us that working from home includes the realities of children, pets and other distractions during online meetings. This learning seems to be applied and acceptable in an online hearing.

> 'Someone could say, "Look, I've got a baby so just so you know they're asleep next door but I might need to go and feed them." But they were able to do that and they weren't really able to do that in a tribunal setting. In comparison to a hearing in person, where people ask, "What happens at a hearing?". You'd say, "You'd have to be there all day." They'd say, "I can't afford to do that." That would be the way they'd fail it. "It wouldn't be possible for me to do it, I can't afford a babysitter. I can't afford to miss all that time off work." Whereas it being remote reduced those issues.'

When asked about the best solution in relation to online or in-person hearings, an interviewee reflected:

> 'In terms of what I think is probably fairest, is that I feel like if either party in a set of proceedings wants the hearing to be in person, they should be allowed to have that hearing in person and then the hearing can either be fully in person or it can be hybrid. We've had a number of hybrid hearings. Say, if one party had problems hearing or for whatever reason weren't comfortable, having it remotely after lockdown but when it was still risky, hybrid hearings worked well because it enabled anyone who wanted to be there physically to be able to be there physically. But if all parties are fine with a remote hearing, I don't actually necessarily see any issue with that.'

Another interviewee, talking us through his experience with the Property Chamber said:

> 'So, you have to have your wits about you, but from an access perspective, how much easier it is for the four of us to be here at 11 o'clock today, than it would've been to tip up somewhere and meet in person. There are definite, I think, advantages but it depends on possibly the scale, the complexity. I think for me, if the insurance, our September hearing, if I'd been asked and given the choice, if I'd been asked, "Would you like to come to court and have an in-person hearing, or would you prefer it to be remote?" I would've asked for it to be remote. Obviously that's just me, but I would've, on that particular instance because of the accessibility.'

Overall, there are benefits for people in joining an online hearing, especially if they are able to choose to do so and possess legal and digital capabilities that equip them with the skills to navigate it. However, technology is necessary for access and use of the online justice space and, as we all know, this can create challenges. These challenges are experienced in different ways by professionals and users.

Challenges and opportunities of technology in justice
Professional perspective

Technology inevitably creates problems, which can be a challenge for all parties in a remote hearing. As one of our interviewees, a housing case handler, shared with us: *"the challenge is obviously for the Property Tribunal moving over to doing remote hearings, getting the platforms up and running, changing over to electronic bundles, and things like that. So, yes, it was a whole sort of change, changing the way of working."*

In the Property Chamber, access to video rooms became available in June 2020 and remote hearings commenced. The Property Chamber President stated:

> 'Once parties had settled into this idea, generally speaking they were content because it was better to them losing the opportunity to have a hearing and they were pleased to be able to engage remotely. However … for those who are aged 50 and below it is more convenient to access an online hearing as they are used to technology. On the other hand, the elderly felt that they were not able to engage on screen and all they had was a phone so they felt marginalized by that.'

The pandemic has shown that certain groups of the population, with varying digital and legal capabilities, are not as able to access the online justice system as others. If there is only an online option to access justice, they are excluded (Finlay 2018). We return to marginalization in relation to technology and access later.

Another interviewee, a housing case worker, commented on the shift to online working during the pandemic and the discussion about digital capabilities:

> 'Zoom suddenly became everybody's favourite video conferencing technology, when nobody had done video conferencing, unless you had relatives in Australia. So that happened overnight, and there are technophobes, but there are real technology zealots in the legal sector, so there was and is quite a tussle still going on about all that. Also, suddenly people did become digitally very literate. But again, I think it's a class thing, in that is something that the middle classes did. I'm not so sure that the working poor and benefits-claimants community engaged in that so much.'

This leads to an important discussion about those who are less able to access the online space. A judge commented:

> 'I think one of the problems is, and again I'm probably being a bit classist here, but people from more deprived backgrounds, and some of them do come

before the tribunal, are typically not as au fait with this computer business. A lot of them don't even have a computer so they use their smartphone. And it becomes apparent that they're looking at everything on a tiny screen and the screen is wobbling all over the place because they haven't put it on a stand. Occasionally, someone has, you know, "my battery has run out, can I just go and get my charger" those sorts of things. In other words, I think that it [online hearings] probably discriminates against people of lower means on a number of levels: understanding the technology, having the equipment, having an environment in which to use it and kind of understanding some of the things that others of us take for granted. … So, yes, I guess there are social obstacles, technological obstacles, spacial obstacles if you want to call it that, but the thing that strikes me is that all of those appear to be unfairly loaded onto particular sectors of particular types of people. Low-income, perhaps not a lot of education, those kinds of things.'

This touches upon a bigger problem in our society, brought to light through the reality of access and non-access to justice observed during the pandemic. For example, there seem to be misconceptions of what the other participants can and cannot see on a screen, informed by one's own experiences. For example, a case worker found that lawyers and judges often think that

'as long as you can see a screen, as long as you can hear everything then we can all do business here. And of course there are real people involved and real people tend to not react in that way and tend to find it quite offensive to have their family members being talked about over video or telephone, and not being part of that.'

The digital exclusion of claimants and appellants is leading to depersonalization in the legal system, whereby individuals are losing their sense of agency and personal connection to the legal process. Furthermore, the 'majesty of the law' is also being depersonalized due to the informal nature of remote tribunal proceedings. While there are potential benefits to this informality, such as increased access and convenience, it raises important questions about the role of the legal system in society and the impact of technology on the administration of justice.

Further, a respondent stated that:

'there is a tendency to think, oh but people do all sorts of things on their smartphones these days, they do their shopping on their smartphone, they organise their holidays on their smartphone. It's an extremely middle-class view, it really is, because it is true, I think that many people have smartphones, but what they use their smartphones for, and how they use their smartphones, is really different. I think the argument that gets made is so disingenuous and

really misses the point. When I see a client, I almost always have to have a look at their phone, because most of them are on Universal Credit, and there is no paper on Universal Credit, it's all done in a journal on someone's phone, so I get to see a wide range of phones that people have. So, it's madness to suggest that we have a population ready to jump into digital dealing, we're just not there yet.'

A SEND judge recalled:

'I had a really difficult hearing where I couldn't see anyone but they could see me. And the problem was I only realized when the registrant basically got very upset at something that I'd said. I would not have said as much as I did if I'd been able to see her face and know that there was a problem emerging. And I thought that was really unfortunate and I think in retrospect perhaps I should have insisted that we get to the point where my connection had been established. I think that visual connection is quite important for that.'

This is a good example of how only listening to a voice and not seeing (even if on a screen) the person's reactions to what is being said can affect the hearing in a negative way.

A Property Chamber case handler noted about technology that it is:

'not that wonderful – unwieldy, parties not familiar with procedures and court proceedings, video conferencing drops, noise interruption, equipment failure, wifi failure, witnesses given promptings from persons not shown on screen, management of parties via screens – not easy, too much staring on screen, eye strain, screen documentation if too large, hard to manage, etc.'

Another Property Chamber case handler commented that:

'the production of digital bundles is not something available to most unrepresented users, even though I didn't have the software to create a bundle despite my day job being a local authority employee with a role of assisting preparation of RRO [rent repayment order] cases. Of course, we probably won't know how many potential applicants have given up an attempt at getting justice/redress when they become aware of the expectations of how a case will need to be prepared.'

In contrast to these accounts, a SEND judge stated:

'I've not had an appellant have any problems. I've had a representative drop off, I've dropped off when my internet's fallen off. You can tell people that are doing it on a phone. It's a lot harder on a phone because you have multiple windows in the same way as Teams, but a decent screen makes life a lot better.'

This account comes from a judge who most likely will have been provided with adequate technology and a reliable internet connection to carry out their role. We need to be mindful of the plethora of challenges someone might have with connecting to the internet, finding a quiet space to attend the hearing, and being aware of how to best conduct themselves and navigate the hearing, to name only a few. In some places, there is assistance available for those who know where to find it. For example, a professional member of the Property Tribunal told us that some councils support tenants who want to make a rent repayment order, for example.

'I think they've supported those applicants and some of these it's like justice for tenants and safer renting, who are charitable type organizations that support tenants as well. So yes, they've generally been helped through the process. In some of the online platforms there are consultation rooms … a person can ask to be put in to have a private conversation with their representative.'

Another account of the online process paints a different picture. An interviewee said:

'My experience is that I felt people had the opportunity to be heard. I've got great admiration for the chairs because that's the role that they play, dealing with it remotely, but making sure that people are still feeling connected with that hearing, and it is a daunting process. If you've never done that you think, it's a legal system, how's it all going to work? The judges and the chairs do explain what's going to happen and the format for the hearing, so they know, and they get the opportunity to put their views across. Certainly, people have appealed, that's the nature of people, but I haven't heard anybody come back to say, "I wasn't allowed to say this", or "I found this difficult to get across", etc.'

Reflecting on the transitions to online processes, a Property Chamber barrister stated that:

'For the most part, I would say that the tribunal has worked pretty well being online. We can talk about the things that are difficult and the challenges, but they appear to be reasonably well resourced, I think they benefit from having a lot of judges who are still practitioners and who are therefore (a) quite used to doing a lot of stuff online anyway and (b) had access to relatively good tech support through their work and they weren't relying upon HMCTS [His Majesty's Courts & Tribunals Service] to do it all. I think certainly that helped a lot.'

Reflecting on the challenges of online hearings, the same Property Chamber barrister stated that it is difficult for unrepresented parties to manage online cases:

'A few things I suppose to go through. It is plainly much much harder for unrepresented parties to manage online cases. I did a significant proportion of my cases where I was against someone who didn't have a lawyer and they'd be doing it on their phone and you can just imagine the difficulty of trying to manage even a modest bundle of a couple of hundred pages on your phone, but if we get into bundles of, I've got one down here that runs to 20,000 pages, doing that on your phone is absurd. That's unusual but not completely out of kilter for what this tribunal does and the difference between the represented parties and the unrepresented parties there is enormous. It's relatively easy for me to buy a second screen so I can have one for the bundle and one for the thing, it's relatively easy for me to just buy a headset. The divide between represented and unrepresented parties got much much greater.'

In some cases, communication among participants at hearings became increasingly difficult online. The Property Chamber barrister stated:

'Secondly, I think you saw a significant advantage linked to that from how you communicate with your professional clients. For example, when I'm in a trial that's online, my solicitor and my clients, we will be in a WhatsApp group and we will be exchanging messages as things go along. It's harder for the lay parties to do that. They may not want to share their WhatsApp details with the three other people who are co-defendants in the block. They may not have a work phone that they can use for that purpose and then just turn it off. I mean I, for example, obviously I'm not going to give my personal phone number to my individual client, they get the work phone and it can get turned off and things like that just are a big difference.'

The Property Chamber barrister also noted the importance of body language in the online space:

'I think as well, to some extent, the unrepresented parties didn't take it quite as seriously when it was online and it's a slightly weird thing to say because often you're fighting about their home and money and such like, but what I mean by that is, when you're all in the room there's that slightly greater sense of decorum because you can pick up on the body language. So when the judge is getting absolutely hacked off with you and wants you to stop, you can sense that. It's much much harder to do online and again the professional parties have an advantage because we are in front of the same judges online again and again and again. You start to learn what to pick up from their online behaviour in a way that, because you're seeing them again and again, you have an advantage in doing. That advantage exists to some degree when it's in person, but it's much less because everyone can pick up on the exasperated huff whereas just watching them, it's harder to do.'

This quote suggests that the emergence of online justice is requiring individuals to develop new capabilities in order to navigate the legal system effectively, such as digital literacy and familiarity with online tools and platforms. This raises the question of whether these new capabilities represent different aspects of legal consciousness, and whether traditional measures of legal capability are sufficient in the context of digital justice. As technology continues to reshape the legal landscape, it is important to consider how these changes are affecting the skills and knowledge necessary to engage with the legal system.

The barrister reflected on the stark differences between types of remote hearings, especially telephone hearings as compared to video hearings:

> 'I would also say there's a real difference between telephone hearings and video hearings. Telephone hearings are the spawn of the devil and should be prohibited because they inevitably lead to much more cross-talking because of course you can't see anything. They give represented parties a significant advantage because the represented party knows to send in a note beforehand of "These are the directions we are seeking" or "These are the key cases I'll be referring to", whereas the unrepresented party has no idea to do that. And I'm not sure you can do much about this, but the telephone system they use for telephone hearings is terrible at cancelling out background noise, you get loads of it, and things like this help because you can manage it a bit more. People genuinely doing telephone hearings on their phone must find it very difficult.'

The data illustrate lived experiences of professionals in online hearings. These experiences differ according to the confidence (and training) the professional has with the system they operate and use as a medium through which to lead their hearing and convey their outcomes. As to be expected, some professionals are more comfortable with the use of technology than others. Those who have encountered challenges during their hearings will have developed methods to address these through which they have consequently altered their attitudes towards and expectations of technology. The data show that professionals have different sensitivities towards the other participants in the online hearing. Some are aware that there are differences in access and ease of use, others assume that everyone has the same access and capabilities, based on their own experience.

Tribunal user perspective

When users were asked about how satisfied they were with communication with the tribunals, one user commented, "The tribunal clerks were very responsive and helped me with queries", another recalled that "sometimes certain issues when time is critical someone to directly speak to would have been better suited", while another said that "given the challenges at home, I am often not *able to answer a call or I miss it. Email only is ideal, plus there is an audit trail".* Another user stated: *"Communication works well in terms of response content, but timescales are a real problem. Phone calls are quicker, but the person answering never has a direct answer. Either they need to go away and check, or they just ask you to keep waiting another few days/weeks."*

Others stated, in relation to the process, that: *"I had time to understand the process, but it is not user-friendly. Instructions are not that clear I don't think. It isn't accessible. For instance, you might struggle if English was your second language."* And some respondents said they were terrified: *"I developed chronic migraines as a result. We were up against the LA by ourselves with our expert witnesses we paid for. These were very basic but fundamental issues about identifying our son's needs and how they should be met."*

Clearly, technology creates added obstacles for some users. A further example of a challenge is the layout of the online hearing. As a SEND user stated: *"Others appeared very small on the screen, I could not see expressions which is a vital part of communication. LA people joined on phone, no video, hiding."*

This situation creates an imbalance in communication and makes it impossible to read facial expressions and catch non–verbal cues. As a Property Chamber case handler told us with regard to the online process, *"There were less benefits to the Tribunal panel, as panel members always benefit from seeing the appellants in person".*

Our interviewees report a range of emotions caused by different aspects of the online process that then get intertwined with what is at stake in their hearing. This adds a new layer to our understanding of digital legal consciousness. All our respondents have reported that emotions form an important part of how they experience and then remember the digital encounter.

An IPSEA employee explained how people struggle with the online process. She describes a parent at a tribunal hearing who had an emotional reaction to a discussion which was not picked up by other participants of the online hearing.

'So, a parent got very upset during a hearing because the nature of what's being discussed can obviously be, you know, emotionally triggering and the parent was asked a question that ended up upsetting them during this virtual hearing and the other participants on the hearing, including the adviser supporter who's sitting at another computer somewhere else in the country, didn't pick up on how upset they were because that sort of thing was, sort of, lost on the screen when we're these little windows. The parent had a friend with them in person to support, not in a legal advice way or an advocacy sort of way, just essentially an emotional support or to help them ask for a break if they needed one, that sort of thing. That person then had to, sort of, step in in a way that you wouldn't really be expecting someone to have to do and to get the tribunal's attention and to explain actually this person is really upset. If they didn't have

that supportive person with them there's a real risk that that could have been missed and that would affect that parent's ability to participate in the hearing because they're there getting upset about a triggering question that has been asked and is obviously not able to focus on the rest of what's going on, and there's a risk there that that would lead to unfairness. And their advocate, their legal advocate, wasn't there in person so they weren't even able to pick up that that had happened. And that's a real difference where you would hope that if you were in the same room in a tribunal hearing situation that it would be obvious and it wouldn't have, you know, ended up going that kind of way.'

Our data show that technology can be an obstacle for those who have a bad internet connection and even *"highly problematic when you've got somebody who's, for instance, profoundly deaf. And you then need a British Sign Language interpreter. Now that throws up a whole new ecosystem of problems"*, a social housing lawyer told us.

'We continue to see a divide between those who are able and those who are less able and disadvantaged to access the online environment. An interviewee [housing lawyer] commented that there is an extent to which all progress leaves some trail of devastation, but surely, we must be minimizing that, surely that's the aim. I'm quite struck by those two paradigms; it is inevitable that court reform and the digital revolution will go hand in hand. But they don't have to, there's no compulsion to shut down courts through digital working.'

This suggests that it is important for people to have the choice to opt for an in-person hearing.

Face-to-face hearings and trust in the legal system

The data showed mixed responses to the question about the benefits and disadvantages of face-to-face hearings. Some think that there is no difference between online and face-to-face hearings, while others suggest that it depends on the type of problem. It also depends on the person bringing the claim and their digital and legal capabilities. As mentioned, there is a strong narrative that emerges from the data about nuanced emotional responses, as recalled by interviewees, as a result of online hearings. This will inform perceptions about the online hearing and be a part of creating digital legal consciousness.

Professional perspective

As seen in the first theme, the judges that we interviewed found comfort and benefits in conducting hearings from home. For example, a SEND judge has strong views on returning to face-to-face hearings:

'It's crap to be quite honest, to say that we work better in a room than on here [online]. We don't, we work better on here [online]. Why throw away all of the progress some chambers have made? In sum, we are judges, we are purely here to get a child with special educational needs into the correct school. We don't need to be in a court. As I say, I just cannot understand why they want to go backwards.'

It may turn out that there might be a future involving hybrid hearings, as well as the claimant choosing what type of hearing they would prefer. A housing case worker stated, when asked about online versus face-to-face hearings:

'I think the tribunal is looking at this, because I think we've realized it does work, and I think some people, bringing cases, they probably feel you haven't got the travel, have you? You can do it in the comfort of your own home. You haven't got to go into a tribunal building, etc., so I think they're looking at certain cases, and I think the rent payment orders are a good example that they could easily be done remotely. I think it's going to be looked at what the parties want, and then the type of cases as well, ones that do fit well with remote and then ones that we want to do in person. We do property inspections as well, or can do, so there may be cases where we want to do that as well and therefore it makes sense to have a face-to-face hearing, so I think it's a bit of a mixture.'

A judge stated that it is most important that the service is consistent, online or in person. They felt strongly that there need not be a difference between online and in-person hearings.

'Just from my experience of remote and face-to-face, I personally don't think it does differ. People still get the same opportunities. It's the same format. It's about the individuals on the panel to make sure that everybody's engaging with it and understanding and feeling that they're getting the opportunity to put their case forward. I think that applies whether it's remote or face-to-face.'

However, when thinking about how the pandemic affected some people's confidence, it might be easier for such people to join a hearing online, although, as the lawyer states in the subsequent quote, people usually have a complex set of issues that are more likely to be teased out in a face-to-face encounter. A lawyer at a law centre told us:

'I certainly have found that a lot of people have become more isolated. They've lost any social confidence that they may have had. So I like to see people face-to-face because I get a good feel of the problems and the issues. They tend to open up more if you see them face-to-face, and you can think, "Okay, we've got a debt problem here, we've got this problem here." And so, you can start

to deal with the case holistically. If you've got somebody on the telephone who's refusing to come and see you because they say, "My social anxiety is too much." And you say, "Well can we do a video call?" "No, I don't want you to see me." It becomes quite difficult. Sometimes the best approach is to see somebody face-to-face.'

Advice providers and tribunals typically have translators and other support to help those people who need assistance with language. As another interviewer points out, if there is a person who does not speak English:

'you are then dependent on the tribunal to engage an independent interpreter of the particular language. That doesn't always flow, it really doesn't flow. And if the interpreter can't access the video hearing properly and then they're on telephone, it's a nightmare. Honestly, it really is difficult. So again, those types of viewings would be better listed face-to-face. The tribunals can be really off-putting and create a handicap to access justice.'

These were a few examples of professionals' views on the benefits and drawbacks of face-to-face hearings. There are, of course, myriad other reasons for and against online and face-to-face hearings, a few of which we will analyse further in the remainder of the book. Now we turn to some user perspectives.

Tribunal user perspective

One of our interviewees reported her experience with the SEND Tribunal during lockdown. She has 'three-and-a-half' children with special education needs.

'I say half because one is in the system at the moment, and two have got EHC plans and during lockdown, obviously we were in lockdown, all the normal activities that my children needed to function quote unquote normally were stopped. So, their routines were totally changed, so I needed support to direct them, how to deal with their anxiety, how to deal with them and it's actually during lockdown that my youngest daughter, her mental health deteriorated because of the lack of consistency, routine, structure. I needed help from the supposed organizations that are supposed to help families like mine.'

This interviewee is aware of help that is out there but not sure where to find it. She has the knowledge of the processes to help her children get EHC plans and has a clear sense that the state or organizations should be there to support her. The challenge was how to access that support for her youngest child during the pandemic. She reflects upon how things were before the pandemic, and the benefit of a face-to-face hearing, in retrospect. The interviewees' encounter with a tribunal dates back ten years and related to a higher-rate mobility claim to get a blue badge:

> '[A]t that point, I had four young children and, in order to get the blue badge back then, you had to have high-rate mobility. So I had to fill out the form for my eldest son, the Department of Work and Pensions turned it down, and then I requested to go to tribunal. I waited 18 months for it, and it was actually as scary as it was because it was shortly after I lost my fourth son, but I was in there for about an hour and a half with my sister and they asked me questions. The Department of Work and Pensions stated, just beforehand, that because it was a complex case, they would have representatives there, no representatives turned up, but the judge and all the people that were in there were phenomenal, and they actually called me back in and apologized for me to have to be there because, if my son was in a wheelchair, it would never have been a case. Off the back of that, they have changed their policy for invisible disabilities, but that was quite a positive experience that I had.'

Our interview data have provided us with a window into the complex discussion about online and offline hearings. We chose to present the themes coming out of our professional data alongside responses from the tribunal user data to show the difference in perceptions and capabilities in order to build an argument for an emerging digital legal consciousness. We do this through observations, in relation to *Type 1* and *Type 2* for this chapter and discuss other types in Chapter 8.

We set out to explore what shapes people's orientation and sensibilities towards the law and technology. We argued that people need a combination of digital and legal capabilities to access and navigate their digital journeys. Together, these form their digital legal consciousness. We found representatives of *Types 1* and *2* (those who are digitally and legally able) in our dataset. There are no clear distinctions between the types because digital and legal capabilities can develop and evolve, as it were. We build our argument for an emerging digital legal consciousness on the foundations of legal consciousness, extended to the online space. Our data revealed that even those who are digitally and legally able in the widest sense can find navigating the online system a challenge. Various interviewees recalled, at different stages of the online process, several *emotional moments*. These emotive responses to digital procedures can interfere with the perception of the process (Chapter 2), as well as boost or diminish confidence in digital and legal capabilities. The range of emotional responses (positive and negative) to the online process forms attitudes towards the online justice system and thereby also forms our digital legal consciousness.

In Chapter 1, we introduced the work of Denvir *et al* (2021) who discussed capabilities through a dataset of judgments and described a psycho-social dimension when analysing their data. It is important to acknowledge what occurs around a person's (digital and legal) capabilities and how this influences them. We also found that emotions have an impact on how people experience and recall their interactions with the online system. Legal consciousness research has now begun to recognize the importance of emotions. For example, Wang finds that:

> a growing trend in legal consciousness research has begun to view the role of one's emotions in relation to the culturally embedded sense of self, giving them more weight in the development of legal consciousness than previously assumed and presenting an entirely different concept of how and when the law may become active in the thoughts and actions of individuals. (Wang 2019: 766; see also Engel and Engel 2010; Abrego 2011; Tungnirun 2018)

Our emotional responses (affective dimensions) are complex and can affect our understanding of reality and subsequent reaction to events (Clore and Huntsinger 2007; Strohminger et al 2011) and we will continue this discussion in Chapters 8 and 9. So far, we can assume that digital legal consciousness can be shaped and reshaped by our emotional responses to the digital journey through the justice system. The shift to online proceedings has highlighted the importance of considering affective dimensions in addition to traditional cost–benefit analysis, as perceptions of fairness and procedural justice play a critical role in the legitimacy of the legal system. This is particularly crucial in the context of online justice, where individuals may feel more depersonalized and disconnected from the legal process, potentially leading to negative perceptions of fairness and procedural justice. As technology continues to reshape the legal landscape, it is important to balance economic efficiency with considerations of affect and fairness in order to maintain public trust and confidence in the legitimacy of the legal system. We discuss the importance of emotional responses to the development of digital legal consciousness and place it in the wider context of this study in the conclusion.

Conclusion

In this chapter we discussed access to the SEND Tribunal and the Property Chamber through our professional and user data, and we made use of the lens of capabilities through which to explore the notion of digital legal consciousness. Four themes emerged from the dataset, and they were discussed from both a professional and a user perspective. This exposed some differences in reported experiences, as well as a number of similarities.

The main similarity was that those who held the online hearings found it much more convenient and timesaving to do this from home. What transpired from the data is that users, especially for SEND cases, found it helpful not to have to travel to a courtroom. Further, the fact that the young person about whom they were making decisions was able to be a part of the hearing, even for just a short period of time, enabled them to put a face to the case they were discussing. Several judges commented how useful that was and several parents/carers also found it useful for the judge to witness for themself the needs of their child.

The stark difference between the parties was the ability and availability of being able to deal with the technical side of the online hearing. Not only did people have a complex set of needs when bringing their cases (whether SEND or housing), but they also had a range of different experiences with the online system. For example, we found that even those users who are digitally and legally savvy still reported various issues when facing the online hearing. Additionally, in refining the interplay between digital and legal capabilities, in the two types of legal consciousness discussed in this chapter, the data provided us with an extra layer through which to understand digital legal consciousness better: namely, through participants' emotions.

In the next chapter we discuss our data in relation to *Types 3* and *4*. These are people who are vulnerable and not easily able to access and navigate the online space.

Notes

[1] We designed and distributed 11 surveys from June 2022 to November 2022: four user surveys (Housing Ombudsman, LGSCO, Property Chamber and SEND Tribunal); four case-handler surveys (Housing Ombudsman, PHSO, Property Chamber, and SEND Tribunal); two judicial and non-judicial panel members surveys (SEND Tribunal judges and judicial and non-judicial members of the Property Chamber); and one for the advice sector. However, despite our efforts to mitigate the low response rate, the final dataset had significant levels of missing data, rendering it unsuitable for our planned analyses. We can, however, use some of the open-ended text responses here.

[2] The ombuds processes are mainly online, so they are not part of this chapter's discussion.

References

Abrego, L.J. (2011) 'Legal consciousness of undocumented Latinos: fear and stigma as barriers to claims-making for first- and 1.5-generation immigrants', *Law & Society Review* 45(2): 337–369. www.jstor.org/stable/23012045

Clore, G.L. and Huntsinger, J.R. (2007) 'How emotions inform judgment and regulate thought', *Trends in Cognitive Sciences* 11(9): 393–399.

Cowan, D. (2004) 'Legal consciousness: some observations', *Modern Law Review* 67(6): 928–958.

Denvir, C., Sutherland, C., Selvarajah, A.D., Balmer, N. and Pleasence, P. (2021) 'Access to Online Courts: Exploring the Relationship between Legal and Digital Capability' (2 May). https://ssrn.com/abstract=3838153 or http://dx.doi.org/10.2139/ssrn.3838153

Elvidge, J. (2018) *Digital Consciousness: A Transformative Vision*, Alresford: John Hunt Publishing.

Engel, D. and Engel, J. (2010) *Tort, Custom, and Karma: Globalization and Legal Consciousness in Thailand*, Stanford, CA: Stanford University Press.

Finlay, A. (2018) 'Preventing digital exclusion from online justice'. Report. London: JUSTICE. https://files.justice.org.uk/wp-content/uploads/2018/06/06170424/Preventing-Digital-Exclusion-from-Online-Justice.pdf

Hertogh, M. (2004) 'A "European" conception of legal consciousness: rediscovering Eugen Ehrlich', *Journal of Law and Society* 31: 457–481. https://doi.org/10.1111/j.1467-6478.2004.00299.x

Merry, S.E. (1990) *Getting Justice and Getting Even: Legal Consciousness Among Working-Class Americans*, Chicago, IL: University of Chicago Press.

Mulcahy, L. and Rowden, E. (2020) *The Democratic Courthouse: A Modern History of Design, Due Process and Dignity*, London: Routledge.

Silbey, S.S. (2005) 'After legal consciousness', *Annual Review of Law and Social Science* 1(1): 323–368.

Strohminger, N., Lewis, R.L. and Meyer, D.E. (2011) 'Divergent effects of different positive emotions on moral judgment', *Cognition* 119(2): 295–300.

Tungnirun, A. (2018) 'Practising on the moon: globalization and legal consciousness of foreign corporate lawyers in Myanmar', *Asian Journal of Law and Society* 5(1): 49–67.

Wang, H.T. (2019) 'Justice, emotion, and belonging: legal consciousness in a Taiwanese family conflict', *Law & Society Review* 53(3): 764–790.

8

Marginalized Groups and Unmet Legal Needs

Introduction

This chapter explores how the pandemic has affected access to advice and redress for marginalized groups. Already marginalized communities are likely to be impacted the most by the pandemic. Yet, we know relatively little about how members of these groups are accessing the justice system, and what can be done to improve their capacity to obtain advice, support and redress. In addressing these questions, the project builds upon, and seeks to extend, existing work about marginalized groups that are alienated by the justice system (Halliday and Morgan 2013; Gill and Creutzfeldt 2018) and whose relationships to authority are characterized by a context of structural disempowerment (Kyprianides et al 2020). Additionally, it is critical to understand people's inaction when faced with a legal problem. Existing research based on legal needs surveys has demonstrated that those experiencing the greatest social and economic disadvantage and marginalization are often the least likely to take any action in response to a rights-based problem (Pleasence et al 2013); in particular, those people who do nothing in response to a problem experienced, which is relatively common in both housing and special educational needs and disabilities (SEND) contexts. This chapter discusses the interview data through the lens of access to justice and trust in justice. However, here, we also apply the lens of digital and legal capabilities (digital legal consciousness) to help understand people's emerging attitudes towards the digital justice system.

By applying the lens of digital and legal capabilities to our data, we were able to tease out the relationship people have with the (online) justice system. Those who are 'digitally assisted, legally unable' (*Type 3*) – for example, the SEND children's parents we interviewed – become alienated from the justice system because they are not able to identify that the problems they are facing might be legal ones and where they can turn to for help. This group is on the margins of being digitally excluded, but if they manage to seek help, as some of our interviewees did, then they become able to obtain assistance in legal and digital access. Although these people have low digital capabilities and low legal capabilities, with assistance, as shown by our interview data, they can access the digital justice system. The homeless people we interviewed, on the other hand, become 'digitally and legally abandoned/excluded' (*Type 4*) in part because of their structural reality of disempowerment. As theorized by Creutzfeldt (2021), our interview data revealed that the homeless people that fall into this category have a limited legal awareness and are not able to access the digital justice system. They have turned their backs on the law, due to the lack of capabilities and complexity associated with assistance. This group has lost trust in justice and thinks that it is just not for them. They suffer the most, as they do not have access to technology beyond that provided by The Connection (a charity supporting clients with access to computers),[1] as well as not knowing if their problems are legal problems or where they can go to seek help (although as our interview data revealed, key workers at The Connection help them navigate support systems). These people are digitally excluded, lacking the skills and confidence to use online technology, and usually not having access to devices or stable internet connections. This is just one layer of social and economic problems that are also related to and constitutive of social exclusion.

As noted in Chapter 1, the term 'vulnerability' is often used to understand the complex nature of the different situations people find themselves in. But defining vulnerability is difficult, not least because vulnerability can stem from external influences, and it also depends on the historical, cultural, social, environmental, political and economic conditions of a given setting. While there are many working definitions, the one most referred to comes from the 1997 report *Who Decides?* (Lord Chancellor's Department 1997). In that report, a *vulnerable adult* is defined as a person: *'who is or may be in need of community care services by reason of mental or other disability, age or illness and who is or may be unable to take care of him or herself, or unable to protect him or herself against significant harm or exploitation'*.

In Chapter 1, we discussed the importance of defining vulnerability in such a way as to provide us with helpful insights into the everyday experience of being vulnerable. Moreover, defining it in this way also helps to build our theoretical framework around vulnerability. Specifically, that a vulnerability approach to the *legal subject* starts from the premise that the concept needs to be rethought and made more representative of the actual human experience:

> It requires that we recognise the ways in which power and privilege are conferred through the operation of societal institutions, relationships, and the creation of social identities, sometimes inequitably. Because law should recognise, respond to, and, perhaps, redirect unjustified

inequality, the critical issue must be whether the balance of power struck by law was warranted. (Fineman 2017: 142; and 2008)

As noted in Chapter 1, McDowell (2018: 104) takes vulnerability theory and focuses on the justice space, in particular, on access to justice. She suggests that:

> utilising these insights, and taking relative privilege, privacy, and autonomy into account, interventions into poor people's courts should seek not merely to provide access to existing legal systems, but also to mitigate the harm caused to low-income people using those systems, foster accountability, and develop meaningful alternatives. This requires a broad approach to providing access, including the provision of opportunities for people to develop the assets necessary for social, legal, and political resilience and change. Attention to functional as well as problematic fragmentations in the state is one way to engage this project and create space for justice as well as access.

It is also important to recognize situational vulnerability in this context, where anyone can be vulnerable depending on the given situation (Dunn et al 2008).

In this chapter, we look at the space created for *justice* and *access* through the lens of people's problems in two distinct areas: housing and SEND. These justice problems create the need for legal advice and legal assistance. Those people who most commonly face difficulties in this area are not typically those who know that they have a legal problem and that there is help available. Therefore, their legal needs are not met. A legal need framing is helpful to understand access to justice, because we have to recognize why people use the justice system in the first place, and, consequently, individual needs cannot be separated from the ability to engage with services. The question of process follows as one way to explore issues of access to (offline and online) justice.

This chapter, therefore, explores how the pandemic has affected access to advice and redress for marginalized groups. Additionally, it is critical to understand the reasons for people's inaction when faced with a legal problem. In addition to users' unwillingness to complain to the ombuds or apply to the tribunal due to the complex practicalities involved in the process and the length of time it takes, people might also struggle to take their complaint forward due to circumstances that deem them vulnerable, such as age, physical or learning disability, physical or mental illness, low literacy, communication difficulties, or changes in circumstances, such as homelessness. One possible approach to understanding vulnerabilities is to distinguish between intrinsic and extrinsic vulnerabilities, and to frame these as continua that operate dynamically depending on the context and situation. This perspective avoids the limitations of a simplistic box-ticking exercise that fails to engage with the complexity and nuance of vulnerability as a concept. Moreover, it also avoids the performative aspect of vulnerability labelling, which can be problematic in categorizing individuals in ways that may be stigmatizing and unhelpful. While the term 'vulnerable' may have negative connotations, it is important to recognize that vulnerabilities are a natural and normal part of the human experience, and to approach them with empathy and understanding rather than judgement or labelling.

The lens of digital and legal capabilities, introduced in Chapter 1, can be applied to our vulnerable interviewees to understand their emerging attitudes towards the digital justice system. Vulnerability can engender a specific type of digital legal consciousness characterized by a heightened awareness of the potential risks and challenges associated with accessing justice through digital means. For example, individuals who are experiencing extrinsic vulnerabilities, such as poverty or social isolation, may have limited access to technology or lack the necessary digital skills to navigate online legal services effectively. On the other hand, individuals experiencing intrinsic vulnerabilities, such as disability or mental health issues, may face additional barriers related to accessibility and accommodations. To address these challenges practically, it is essential to adopt a user-centred approach to designing digital legal services that takes into account the needs and perspectives of individuals with diverse vulnerabilities. This may involve providing targeted training and support to help individuals develop the digital skills needed to access legal services online, as well as ensuring that online legal services are designed with accessibility in mind. Additionally, it may be helpful to establish partnerships between legal service providers and community organizations to provide targeted outreach and support to vulnerable populations. By taking these practical steps, we can work towards creating a more inclusive and equitable digital legal system that meets the needs of all individuals, regardless of their vulnerabilities.

With the help of these general types, we start to unpack the complex nature of a multidimensional understanding of access to digital justice from the perspective of the user. We identified our user responses in Chapter 7 to be a combination of *Type 1* and *Type 2*. In this chapter, we identify our user responses to be a combination of *Type 3* (*digitally assisted, legally able*) and *Type 4* (*digitally and legally abandoned/excluded*).

The types do not have well-defined boundaries in the data, just like in real life. For instance, a person classified as *Type 4* could participate in a tribunal with the help of a representative or supporter: our dataset did not include an example of this. There are also more intricate differences within and between the four types, such as variations in digital and legal capabilities that may change. For our study, the interview extracts that now follow are discussed in the context of building our evidence base for developing a notion of digital legal consciousness.

Our interviews, especially with SEND users and the advice sector, produced a strong narrative of disabilities and hidden vulnerabilities in relation to online hearings. For example, an Independent Provider of Special Education Advice (IPSEA) employee stated:

'But in practical terms it's just picking up again on that emotional side of things. There's also other things that are unseen and unheard on the phone or via Teams or Zoom or whatever it might be, the way that someone's providing support or advising someone. Which can make it more challenging. I mean, some parents are very open and they say "Oh no, I've got real difficulty with numbers", or "I haven't been diagnosed but I think I might also have ADHD [attention-deficit/hyperactivity disorder] or whatever so I do find it difficult to organise myself". But others will just not have engaged in anything like this before, not have possibly even sought professional advice in anything, let alone in the context of something that appears pretty daunting to most people. So in terms of providing a service which offers advice and case work support, our volunteers have had to keep in mind, as they would have done beforehand as well but maybe more so now that they're literally never going to meet the person face-to-face. There may be hidden difficulties that don't come across as well.'

We draw on the in-depth interviews conducted with six parents of SEND children and seven homeless people engaging with The Connection, a charity that supports people experiencing homelessness. Interview questions revolved around the eight steps that we identified which users go through when seeking help (see Chapter 4 for housing and Chapter 5 for SEND). Interviewees were asked to share their stories, including questions around whether they had experienced any housing/SEND issues during the pandemic, at what point they became aware that there was a problem, and how they went about addressing that problem. Next interviewees were asked a series of questions on taking action, including whether they had tried to get support for their issues, how they looked for services, and whether they experienced any difficulties knowing how and where to look for help. Participants were also asked about the advice sector and any support or guidance they had received before being asked to reflect on their experience of any intermediate processes involving their landlord/housing association in the case of housing or any organization involved (for example, local authority, school or governing body) in the case of SEND. Those participants that had contacted a tribunal or ombuds were asked an additional set of questions revolving around how they went about accessing the justice system, which institutions they had approached, and how much time they spent trying to sort out their problem before approaching the institution, as well as any expectations they had. Next, they were asked to reflect on their experience of engaging with the institution, including what worked well, what barriers they had faced, and the extent to which they trusted the process. Finally, participants were asked what they thought could have been done during and after the pandemic to improve users' capacity to obtain advice, support and redress.

Housing

Perhaps unsurprisingly, none of the participants recruited via The Connection had contacted a tribunal or ombuds (*Type 4*) and so were not asked any additional questions regarding the processes associated with accessing, engaging with, and receiving service from these institutions. Consequently, this section will focus on the themes that are important to them, and particularly on the importance of intermediaries and charities. At the end of Chapter 4, we discussed an example of a tribunal user and highlighted the importance of charities and intermediaries. In that example, the person was digitally and legally able but found the assistance of these organizations of great importance to her navigation of the system to get support. In the following we present some responses from those participants that we deemed vulnerable (as outlined in our definition of vulnerability above) sorted into five categories that emerged from the interview data: overcoming barriers to access to justice; providing support and guidance for legal needs; taking action for justice; navigating the impact of the pandemic on access to justice; and promoting knowledge and awareness of tribunals and ombuds.

Overcoming barriers to access to justice
Structural disempowerment: marginalization and stigmatization
All participants reported being homeless for long periods of time, living on the streets and sleeping in tents. They had all spent different amounts of time living like this. A harsh psychological and material reality was evidenced by participants' stories about their housing issues.

For example, one participant has been on-and-off homeless for about eight or nine years and in and out of hostels during this time. He experienced London's free hostels for the homeless as instigators for drugs and drink, and so he left and had nowhere to go.

'*The hostel life is pretty bad, drugs and drink. I don't want any of that anymore, so they put me in a place where there's going to be drinkers you're going to follow. You're just going to follow it because it happens all of the time. You*

179

think it won't happen to you, but your mate next door gets a bit of money and then it all happens again. So, I walked out. … I had to go.'

And another respondent is currently living in a tent because of the damage in her flat that forced her to leave.

'My flat everything broke in it basically. The shower was leaking, the toilet pipe broke off, the floor was sinking in, the kitchen sink was broken, the lights were broken, the electrics were broken. Basically, it got condemned and I couldn't get another flat. So, I moved here. Just living in a tent. I've lost a lot of my stuff as well. Just waiting to get help at the moment.'

Moreover, this group of participants appeared to struggle with physical and mental health conditions on top of this. For example, one of the researchers handed an interviewee the participant information sheet to read; the participant appeared embarrassed and quickly put it away without attempting to read it.[2] This relates to the wider issue of literacy in accessing justice for this group. As we will come to see, this is illustrative of the ways in which this vulnerable group is treated: people just do not 'clock' things that are significant to this group (for example, that they might not be able to read).

Providing support and guidance for legal needs

The people we spoke to 'took action' by going to The Connection, to seek support in resolving their housing issues. Participants found out about The Connection via word of mouth (friends mostly). All participants spoke very positively about The Connection. They explained that The Connection offers them food, temporary shelter and a shower. The Connection also advises its clients to seek help from the council to get temporary accommodation. One interviewee said:

'This place makes sense. I like this place. They've saved me a few times when I really did need them. These people opened the door, fed me, shower. I've got a lot of good things to say about Connections, they're good people … they are helpful. They put me in touch with St Mungo's and the outreach and they keep ringing us. We've been in touch with Camden council and said we're looking for temporary accommodation.'

Taking action for justice

Three out of seven participants had engaged with the council, following advice from The Connection key workers. They described their experiences to us. One man tried to get support from the council, but the council was not helpful. He felt discriminated against due to his housing status, age and socioeconomic status. This man also struggles with literacy skills so engaging with the council was difficult:

> 'The council just gave me a couple of letters and said, "That might help you. Can you read?" I said, "Not really". At the time I couldn't read that well so I didn't understand the sheets of paper she was giving me anyway. … So I've been down that route and I've been in and out of the council places, numerous times. It just doesn't make sense to sit there. It's all expensive, no one can afford it. Or they tell you to "Take that to the job centre" and because of your age, you are a lot younger than what you are now, they go, "Oh priority, priority". "How old are you?" You tell them you're 17, they love you. "I'll make a couple of phone calls".'

A female interviewee was advised to seek help from the council via St Mungo's outreach to get temporary accommodation for herself and her partner. Although she explained that she feels that things are moving and that she is being informed about what is happening, St Mungo's/the council have not told her how long the process will take to get her the support she needs. Another male client, who is currently residing illegally in the UK without a visa, was put in touch with immigration services by The Connection to get him the support that he needs. Again, he was unsure about how long the process will take.

Navigating the impact of the pandemic on access to justice

Participants described how the pandemic negatively impacted on their housing issues and mental health:

> 'Mental health as well. It puts you in a place where you don't know where to go if this place is shut and they're scared to touch you or scared to come within this thing. It was worse for us because we're stuck out here and then people don't want to be near people outside. I had all of my vaccinations, kept my little thing "I've had mine mate, don't muck about".'

Although the pandemic resulted in homeless people being housed in temporary accommodation, one man explained that the pandemic worsened his housing issues as it disrupted access to homeless services:

> 'Pandemic, they were offering people in hostels temporary Travelodge, and other ones-, you know them yourself Miss-, and there were other places. That's all well and good sticking someone there temporarily but they're going to think, "I can get quite used to this", then all of a sudden, you're going to chuck them back outside. It was hard. The government, whoever it was, was closing the

toilets and that. … I'm not going to lie to you, the only thing you weren't getting is your full breakfast on the table because you're all close to each other, so they put a stop to that too.'

Promoting knowledge and awareness of tribunals and ombuds

Most participants had not even heard about a tribunal or ombuds, and where participants had heard of them, they struggled to understand what they do: *"I've heard it yes but understanding what they do there is new to me. I wouldn't know where to start. Tribunal sounds like a council's worst nightmare. That's what it sounds like to me. Is that what it is? I've got a pain in my chest just saying that."*

Reflections from The Connection staff

We also spoke with some of The Connection staff. They described their roles: *"We are part of the resource centre team, basically the first team that anyone meets when they come in and we get them in the system, we show them around the services and we're helping with the services as well"* (resource manager).

According to The Connection resource manager and the key worker we spoke with, seeking temporary accommodation is the most common issue they deal with in relation to housing, closely followed by immigration and benefits (needing help to claim Universal Credit), and clients are advised to engage with the Department for Work and Pensions (DWP), the council, Citizens Advice, legal advice services or the police to resolve these housing issues. The resource manager and the key worker explained the process clients at The Connection undergo to engage with the council to help them resolve their housing issues. The resource manager reported that this route through the council for their clients to find temporary accommodation proves difficult for them for the various reasons we outline in the next section. A DWP worker therefore attends The Connection once a week to support clients with aspects of the process such as the paperwork involved.

The resource manager and the key worker explained why support routes are inaccessible to their clients, highlighting certain obstacles/challenges their clients face when trying to secure temporary accommodation. As noted by clients themselves, these obstacles/challenges included marginalization/ stigmatization including mental/physical health impairments and wider issues among this group, such as literacy. However, the biggest issues relate to council prioritization processes, clients not having a local connection to Westminster, and the limited housing options available.

Clients often do not have a local connection to Westminster (the area the charity serves) which complicates this process:

'So, the main issue that we have is a lot of clients don't have a local connection to Westminster or to any borough pretty much a lot of the time. So, a local connection is where you've been living out of three of the last five years ... where you have a local connection it means there will be services who you'll be eligible to be helped by. So, finding out where someone's local connection is vital if they'll be eligible for certain types of support. Often, people may have a local connection somewhere else but they're inclined to go or access things in various areas, for reasons, and then it narrows down the options for them. So, if my local connection is Islington but I'm refusing to go there, then I'm left with private rented accommodation. You can't just go to any council you want and be assessed for housing, because you have to have a local connection to the borough. So, you have people, say, from Liverpool come to London, "I want to work and live here", and can't approach any council in London because your local connection is Liverpool. So, your options are pretty much limited.' (Key worker)

Moreover, the council prioritizes certain clients over others:

'Everybody has different circumstances, the council obviously put some people on priorities and the other ones put them at the bottom of the list I guess. That depends on what the situation really is, we can do nothing about that and their priorities. ... I guess the problems could be if they're entitled to get help or not, some people that are here are especially vulnerable but sometimes they cannot get help. I mean we have certain resources and we utilize them in the best possible way really, but then after that it's about eligibility from the rules, the government I guess, the council ... you can do up to a certain point, help up to a certain point, then it's not up to us anymore.' (Resource manager)

The key worker explained that being homeless alone is not categorized as a complex need. Instead, having psychological, mental or physical health problems on top of experiencing homelessness would deem someone 'vulnerable enough' to be prioritized.

Those with less complex needs, not prioritized by the council, are left with few and expensive housing options, due to the lack of available housing:

'But then the issue with that also is if you don't have any real complex needs you may not be deemed as a priority by the council to be placed into immediate accommodation, and then we have to then start looking at the private rented route as well, because there is just a lack of housing in general.' (Key worker)

Landlords also frequently refuse tenants that are on Universal Credit and/ or housing benefits, further toughening the process. Moreover, to be able to afford private rented accommodation, then, clients usually have to go through a process of securing employment in the first instance and sleeping rough until they receive their first paycheck:

> *'Yes, the lack of accommodation in general, finding employment and sleeping rough whilst trying to work, finding suitable accommodation for a decent price. Private rented accommodation is often very expensive, as well, so just the cost of living. … Yes, because it's inner London as well, primarily for finding people who are working a minimum wage job, and you're in zones 1 to 4, privately rented, it is just very expensive, and it's very hard for them. So, finding suitable accommodation, being able to travel, being able to sustain themselves, is very difficult for people to do.'* (Key worker)

It was clear what *would* help their clients access the system: more available housing. Both the key worker and the resource manager reported that affordable accommodation, including hubs and shelters, would assist vulnerable people out on the street to move into accommodation willingly.

The resource manager and the key worker also reflected on the impact of the pandemic at The Connection. During the peak of the pandemic The Connection was closed, and everybody was temporarily housed in hotels. However, The Connection played a role in this process, housing people in hotels/hostels and providing services. A key change following the COVID-19-related lockdown was that The Connection changed from being a 24-hour service to being a day centre only. Council applications also became available as an online process as a result of the pandemic:

> *'So homeless applications made by the council, primarily done online but can be done face-to-face once you get an appointment with-, different councils vary in their process, especially after COVID as well. It used to be a process where you'd turn up at your local council housing options and you would be there for quite a few hours, wait to be seen, and do the application with a housing officer. They would then look at the application, see whether you're priority needs or not, or whether you will be eligible for temporary accommodation, and then they would be able to, kind of, go forward with that. … It is all now done online. They are still face-to-face, but now it, kind of, balances out, it's not all one way.'* (Key worker)

The extreme marginalization of our homeless participants illustrated that this is a vulnerable group with little to no access to justice, even limited access to the council. They represent *Type 4* in our framework: those people who need the system to work for them, but cannot reach it.

We also spoke with Age UK, a charity that supports elderly people in a variety of daily matters – one of them, relevant to our discussion, is digital exclusion. For example, Age UK's digital inclusion programme can provide older people with the digital skills necessary to live more confident, independent lives.

An Age UK employee explained that the programme is designed to support the elderly:

> *'We work on the same programme which is Age UK's biggest ever digital inclusion programme spanning the next, sort of, four years, and working with 50 different local Age UK organizations. Age UK has quite a long history of digital inclusion projects, and the model that we're working with now is tested and proven to the extent that it can be. It's a volunteer led programme, so, the idea is that we as a central organization provide funding to the local Age UK partners and within their specific area they will recruit volunteers who will support older people with how to get online, how to use technology, whatever it is they want to be able to do. It is very personal led, there's no set lesson plan. It's all designed in conjunction between the volunteer and the older person accessing the service. So, they will come with a diverse range of needs, thoughts, wants, whatever it might be that they want to learn about the digital space.'*

COVID-19 disrupted the running of the programme, and it was adapted accordingly:

> *'[T]he model itself was developed on very much an in-person delivered service. Obviously, with COVID and the pandemic, there were a lot of adjustments and alterations made to ensure the programme could be delivered more flexibly. So, including online delivery and in person, and at the moment we're taking forward that hybrid model. It's a real mixture, again depending on the older person's needs and also circumstances, it might be they prefer an online connection rather than leaving the house and going to a community centre, conversely it might be that they prefer a group session where they get the social aspect of meeting other people as well. So, it's very very tailored to the individual.'*

In sum, the efforts made to work towards digital inclusion are evident for the more vulnerable groups of people in our society, such as the street population; charitable organizations work hard to help them get their voice heard, and support them with the practicalities involved in navigating, in this context, the housing system. However, that is often not enough to get those more vulnerable the support that they need due to obstacles/challenges such as marginalization/stigmatization, mental/physical health impairments, and wider issues such as literacy. The professionals that we spoke with suggested that efforts need to be tailored to help people overcome their own barriers, whether that is around confidence, cost, access or skills.

We next look at SEND users – another vulnerable group, albeit experiencing different vulnerability from homeless people, but for whom access to justice is still difficult.

SEND

Participants in the SEND context had contacted a tribunal or ombuds and so were asked additional questions regarding the processes associated with accessing, engaging with, and receiving service from these institutions. Consequently, this section will focus on the themes that are important to them, particularly on their experience of engaging with, and receiving service from, various advice organizations, including the importance of intermediaries and charities.

In the following section, we present responses from those participants that we deemed vulnerable (in line with the theme of this chapter), sorted into four categories: access to justice for vulnerable populations; legal processes affecting access to justice; negative perceptions of legal services; and positive perceptions of legal services.

Access to justice for vulnerable populations

Participants/parents indicated feelings of helplessness, especially during the pandemic. Often these feelings of helplessness can lead to depression among parents/carers:

'To be honest at the time my mental health was through the roof anyway. Only surface level because of how I was struggling myself personally, and then obviously trying to keep as calm a ship as possible in my household. … Today or tomorrow if I was having an issue, I would but obviously in the height of the pandemic, four children with autism at home, home schooling them. My mind couldn't, I'm neurodiverse myself so my thought process was just do what I can do to get through today.'

Participants, overall, did not feel that 'the system' is there to support vulnerable families like their own:

'I have to make myself vulnerable for them to listen to me. I struggle with mental health myself personally, I have an ADHD diagnosis, but it's not something I like to use as, "Well I'm depressed", just to get help. I shouldn't have to use my mental health or my neurodiversity to get the help that is technically there. I actually got told last September that I should be grateful that I'm getting transport for my two sons, and I tell you, I literally hit the roof. I lost my mind, and I'm not an aggressive or an angry person, but I just had a go at him, and

I said, "Do you think I chose for my children to all be autistic? Do you think it's what I chose for my life to be? How dare you tell me I should be grateful. No, if you want to use that, you should be grateful that you're in the job that you're in because it's my children why you're in the job that you're in", and I don't think, and one thing I said from day one, is a lot of these people are in these jobs that, at 5 o'clock, they finish their job and they go home. My life doesn't stop, it's 24/7, and my children are just a file on their desk. They don't have to take me or my family seriously.'

Some felt that future efforts are unlikely to work to support them and their children with SEND issues:

'I think, personally, the government. Not local government, national government. I really do feel there are not enough people in the actual government who understand what parents of children with additional needs go through on a daily basis. I don't think there's enough emphasis on our lives. Yes, we've got the Paddy McGuiness and we've got Katie Price but, with all due respect, they have money, so everything they need they can get at the click of a finger whereas, like myself, my daughter, after four times, has now gone through assessment. We would've had to privately get her assessed if we couldn't go through the local authority.'

This first theme apparent in our data highlighted the vulnerability and helplessness experienced by those seeking to navigate the system; leading to help-seekers feeling that 'the system is working against them' rather than 'with them' and feeling pessimistic about receiving the support that they need.

Legal processes affecting access to justice

Parents shared their children's SEND issues with us, highlighting the complexity of these issues. For example:

'Well, it was generalized and specific. So, no one particular problem. It was just generally. My children, like I say, two of my kids have got EHC plans. They should have been in school, but I was made aware that, no, they can't be in school. It was only at a crisis point for myself when I nearly had a breakdown, when somebody-, I literally cried down the phone to someone and then they were like, "Well, actually, your children who have got an EHC plan should be at school". And suddenly they're in school. So, a crisis has to happen before anybody picks up a baton and does what they need to do. I think very much in the borough that I live in, Lewisham, that's what it is generally. A crisis has to happen before anybody takes action. So, they wait until blood is drawn and legs are broken before they actually pick up and say,

"Actually, yes, maybe". But before you're at that point I'm screaming and shouting, and screaming and shouting, and nobody hears me. I have to cut my arm and expose my bone before you say, "Oh, actually. Yes, I can help you". That's the general consensus of my life in the last 13 years of having numerous children with special needs.'

Due to the complexity of their SEND problems, parents reported spending a lot of time trying to resolve the problem before seeking help. For example: *"A long time. The whole first lockdown. I would say about six, seven months before I actually stepped up and tried to start getting professionals in. Yes."*

Reflecting on the organizations they have gone on to seek help from, participants perceived the school and local authority negatively, but certain advice organizations and the SEND Tribunal positively.

In sum, our interviewees recalled the time spent trying to resolve their complex issues before seeking help. Those they engaged with in this initial process were sometimes helpful – in the case of specialized advice services – but also unhelpful – in the case of the council and schools.

Negative perceptions of legal services

The local authorities and their children's schools posed significant access issues for parents during the pandemic. Talking about her experience trying to get help from the local authority, one participant explained that even accessing the local authority was difficult:

'Well, I was a trustee for Luton Parent Carer Forum, and then obviously through them trying to get the relevant departments within the local authority to support, and as a trustee it was my role as well to support other parents, but obviously if I couldn't get support for myself, there's no way I could have got support for the families. … Well, ironically, I used to actually be a trustee for an organization that was supposed to help, but even being a trustee of that organization we weren't being contacted, or we weren't being supported by the local authority who were supposed to support us, to support families anyway. But yes, we couldn't get hold of anybody, everybody was obviously working from home, but the phone numbers that we had were their office numbers, and it wasn't being diverted and nobody could actually help in the way that the help was needed. So we were left on deaf ears, basically.'

Talking about her experience trying to get help from her child's school, one participant said:

'Well, I did try to go through my children's schools. Again it was quite difficult because one of my children's schools is an independent specialist school, so they

don't have the local authority on tap because they're independent, and the other school although they do have links with the LA, they had the same issues that I had. Not being able to get hold of or contact the relevant departments to let us know what the situations are. ... Yes, the school was supportive, but again, they could only do what they could do with the remit that they had.'

Post-pandemic, accessing the local authority did not become any easier:

'Not really with the local authority and I stepped down as a trustee. ... So no. Not really. I'm lucky that I don't have a paid job and my husband is very supportive, and I can talk and I can speak, and I can shout loud. There are so many families that can't, who are just being missed or ignored, or hoping that LA whether it be the SEN team, whether it be the social worker for the children, whether it be. They're banking on the naivety of the parents who don't know. So, they gaslight them to make them believe that they aren't entitled to A, B, C, D and E, or they're not entitled to it or they don't get told what they are entitled to. So, the local authorities bank on people, families and parents, not knowing.'

In fact, participants described negative encounters with the local authority:

'To be honest, it's tiring. As I said, it's correcting people on doing their job in the appropriate way, it's simple, basic communication. For example, on the 27th of July, which was my 44th birthday, I had an email from the Transport Team basically saying, "Because you didn't reapply when we sent you a letter on the 7th of April, transport for your two oldest children ... has been stopped". ... Like I said, about ten years ago, I made my first formal complaint and, every three or four months since then, I make complaints with the local authority, it goes on deaf ears, nobody actually cares or it just slips through nets. All I get is an apology. ...

Not communicating, the case worker who has my child's EHC plan, not communicating, they put more of the onus on the school, say, "Oh, the school should have it, the school should have it". They don't update the parents. For instance, my son who's just left primary school, he got diagnosed at 2, he had a little bit of intervention when he was in reception, and nothing. ... So when he got into Year 5, I requested for him to have an updated speech and language assessment. The local authority requested that ... it got turned down. And I pushed, I pushed, but on their behalf they've tried as much as they possibly can, but with their resources, there's not much that schools can do, and now with my daughter, who is, like I said, getting a diagnosis, and again, she's in school, she's learning, but she doesn't function in unstructured time. So my next battle, and it will be a battle, is to get my daughter an EHC plan, which I know will be a battle.'

In sum, negative interactions with key stakeholders, such as schools and the local authority, had an emotional impact on help-seekers that was the cause of psychological stress which further worsened their ability to access the system.

Positive perceptions of legal services

However, there were a number of services where participants reported positive experiences, including three advice organizations, and the SEND Tribunal.
The first is IPSEA:

> 'I had to give up my job to look after my daughter as she was out of school so thus was a challenge but also an opportunity as with the help of IPSEA advice I was targeted and strategic in my approach. My criticisms were that CAMHS [child and adolescent mental health services], LA and social care were not working together to progress a reasonable plan. I made sure I got very good evidence to support the school placement. I rang professionals and challenged them from an evidence base.' (Survey data)[3]

The second advice organization is the Drumbeat outreach team,[4] who attend schools to support parents by providing a range of courses for parents, as well as reports for the schools on how to support the children concerned. The third is Lewisham Autism Support,[5] an organization that supports families with children suffering from autism spectrum disorder (ASD), attention-deficit/hyperactivity disorder (ADHD) and other neurological disabilities.

Experiences with the SEND Tribunal were positive. One participant who went to the tribunal for one of her son's Disability Living Allowance said it was an *"exceptionally great experience"*.

In contrast, a few participants reported negative experiences at the SEND Tribunal. One participant said: *"the tribunal never read the full case, they were clearly biased and never took the critical mental health state into consideration. They were rude, disrespectful, and even my solicitor said the panel decisions were heartless and unfair with their conduct"* (survey data).

In sum, these excerpts from our data show how complex people's interactions on their digital journeys can be. Help-seekers appear to go through a roller-coaster of emotions during the complex process of finding the right support, even when their interactions with organizations are deemed positive experiences overall.

Conclusion

This chapter has explored how marginalized groups – specifically homeless people and disadvantaged parents of SEND children – struggle to access the justice system to get support for their housing or children's SEND issues. We also explored how the pandemic has affected access to advice and redress for these marginalized groups and reflected on their perceptions of what could have been done during and what could be done after the pandemic to improve their capacity to obtain advice, support and redress.

In Chapters 6 and 7 we explored how individuals navigate the online system, which unveiled a multifaceted landscape. While digital and legal capabilities play a role in shaping one's digital journey, our findings indicate that other factors, such as emotions, temporal vulnerability and hidden vulnerabilities also exert a significant influence. In this chapter, we have acknowledged the usefulness of the digital and legal capabilities framework in understanding access to justice. However, we stress the importance of considering additional elements such as procedural justice, emotions and both temporal and hidden vulnerabilities in understanding better who accesses the justice system. While the notion of digital legal consciousness may offer a starting point to dissect these intricate interactions, it is crucial to recognize that any organizing theory/concept must be fluid and adaptive to the constantly evolving nature of the justice system. The next chapter brings all our research findings together and proposes a model through which to understand (non) access to online justice.

Notes

[1] www.connection-at-stmartins.org.uk
[2] The researchers then explained to the interviewee everything that was written on the participant information sheet about the research and about their right to opt-out at any time.
[3] We designed and distributed 11 surveys from June 2022 to November 2022. However, despite our efforts to mitigate the low response rate, the final dataset had significant levels of missing data, rendering it unsuitable for our planned analyses. We use here some of the open-text responses.
[4] The Drumbeat Outreach team are a team of members of staff who go into schools and who support parents in Lewisham. They offer a large range of courses for parents and, if the school supports them, they also go into the school to give the school a thorough report of how to support that child.
[5] There is a service called Lewisham Autism Support. There are only three of them, and they are supposed to see and support the family of every single child that gets a diagnosis of ASD, ADHD or a neurological disability.

References

Creutzfeldt, N. (2021) 'Towards a digital legal consciousness?' *European Journal of Law and Technology* 12(3).

Dunn, M., Clare, I. and Holland, A. (2008) 'To empower or to protect? Constructing the "vulnerable adult" in English law and public policy', *Legal Studies* 28(2): 234–253. doi:10.1111/j.1748-121X.2008.00085.x

Fineman, M. (2008) 'The vulnerable subject: anchoring equality in the human condition', *Yale Journal of Law and Feminism* 20(1).

Fineman, M. (2017) 'Vulnerability and inevitable inequality', *Oslo Law Review* 4(3): 133–149.

Gill, C. and Creutzfeldt, N. (2018) 'The "ombuds watchers": collective dissent and legal protest amongst users of public services ombuds', *Social and Legal Studies* 27(3): 367–388.

Halliday, S. and Morgan, B. (2013) 'I fought the law and the law won? Legal consciousness and the critical imagination', *Current Legal Problems* 66(1): 1–32.

Kyprianides, A., Stott, C. and Bradford, B. (2020) '"Playing the game": power, authority and procedural justice in interactions between police and homeless people in London', *British Journal of Criminology* 61: 670–689.

Lord Chancellor's Department (1997) Who Decides? Making Decisions on Behalf of Mentally Incapacitated Adults'. http://webarchive.nationalarchi ves.gov.uk/+/http:/www.dca.gov.uk/menincap/meninfr.htm

McDowell, E.L. (2018) 'Vulnerability, access to justice, and the fragmented state', *Michigan Journal of Race and Law* 23(1&2): 51–104.

Pleasence, P.T., Balmer, N.J. and Sandefur, R.L. (2013) 'Paths to justice: a past, present and future roadmap'. Report. UCL Centre for Empirical Legal Studies.

9

Conclusion

Introduction

In this final chapter we discuss our empirical findings through the theoretical lenses introduced in Chapters 1 and 2. Our focus is on the user (help-seeker), how they access online processes and what shapes their perceptions of these processes. We set out to argue for a broader understanding of access to justice, to include non-legal actors, and to take into consideration vulnerabilities; ensuring access to justice extends to all members of society, not just those who are legally and technically capable. The current access to justice framework, which is heavily focused on legal tradition and legal confines, must evolve to be more inclusive and consider the needs of non-expert individuals. For example, the traditional design of courthouses has not considered the needs of lay users (Mulcahy and Rowden 2020), and as we move towards an increasingly digitalized justice system, it is crucial that access to justice is safeguarded for all (Denvir and Selvarajah 2022; Mentovich et al 2023). Online justice presents both opportunities and challenges (Schmitz 2019), and it is vital that we proactively address these to ensure that the justice system is accessible to all. Moreover, access to justice is a personal journey, and it is imperative to recognize that individuals may have unique emotional responses that cannot be solely attributed to their level of digital capabilities. The distinctions between in-person and online experiences further complicate this process, as emotions can vary greatly in each scenario. The online setting can prompt emotional responses, even if one has support from a lawyer or an advocate (Hou et al 2017).

We advanced the theoretical scholarship on, and methodological approach to, legal consciousness through the lenses of digital legal consciousness (Creutzfeldt 2021) and procedural justice (Creutzfeldt and Bradford 2016) to secure a better understanding of what the online system expects of people; what people expect from the online system, and how they make sense of it. We discuss this in four parts. First, we propose a model that better represents an inclusive idea of access to (online) justice. We do this by bringing theories from different disciplines together to think about digital journeys.

Digital journeys: revisiting digital legal consciousness

We introduced the notion of digital legal consciousness to understand those people who access – and those who do not access – the online justice system. We understand digital legal consciousness as the combination of the digital and legal capabilities people need in order to navigate the online space, to enable their digital journeys (Denvir et al 2021; Creutzfeldt 2021). Our data showed that the original four types (made up of a combination of different levels of legal and digital capabilities) are more complex, can overlap and can be transient or visible. Our data helped us develop a more nuanced notion of digital legal consciousness.

Our findings indicate that a digital and legal capabilities framework is a valuable tool, but there are additional layers of complexity when exploring access to justice, such as trust in administrative justice and procedural justice in interactions (Chapter 2); navigating the divide between official guidelines and help-seeker journey experiences (Part II); accessing digital justice in a post COVID-19 administrative justice system (AJS) landscape (Chapters 6 and 7); and how members of marginalized groups access the justice system (Chapter 8). We uncover emotions, temporal and hidden vulnerabilities, and the interplay between online and offline experiences, that must be considered.

Our research has shown that the lived experience of people's digital journeys is complex and multifaceted. Our data, therefore, exposed themes that transcend and connect the four types, usually blurring their boundaries. These themes need to be woven into our understanding to help us think about access to justice, digital journeys and face-to-face pathways to justice. The shift to digital spaces presents a unique set of challenges that must be addressed if justice is to be truly accessible to all.

The four categories of digital legal consciousness (*Type 1*: Digitally and legally literate/enabled; *Type 2*: Digitally agnostic and legally able; *Type 3*: Digitally assisted and legally unable; and *Type 4*: Digitally and legally abandoned/excluded; Creutzfeldt 2021), serve as a useful framework for considering the varying levels of legal literacy, digital literacy, access to technology, and social and cultural capital that exist among individuals. However, it is important to note that these categories are not always clear-cut and may not capture the full spectrum of experiences. There may be individuals who exhibit characteristics of both *Type 1* and *Type 2* service users. They may be digitally and legally literate in some areas but lack knowledge in others. Similarly, *Type 3* service users may have access to digital resources but still struggle to navigate the complex legal system. Additionally, the framework may not account for other factors, such as language barriers, disability or socioeconomic status, which can impact an individual's ability to participate in the justice system. Therefore, it is essential to move beyond these four categories and explore other factors that may influence an individual's digital legal consciousness. This could include factors such as community resources, trust in legal institutions and experiences with discrimination or bias. Further, the digital systems themselves often pose problems to the users as they are not very intuitive or user-friendly.

Our research indicates that, when procedural justice is present in online processes, it enhances the legitimacy of those processes. Furthermore, we discovered a connection between procedural justice and digital legal consciousness, which is influenced by three dimensions as observed in our data. These are affective, digital and compound dimensions (see Figure 9.1). Moreover, our findings suggest that the online process itself can produce dimensions that help us understand it better. The *affective dimension*, including procedural justice and uncertainty management, is crucial to understanding an individual's experience with the legal system. This dimension encompasses feelings of fairness, respect and trust in legal institutions. The ability to manage uncertainty and communicate effectively can also impact an individual's ability to navigate legal processes. Moreover, the *digital dimension* must be considered when examining an individual's

Figure 9.1: Digital journeys: a more nuanced understanding of digital legal consciousness

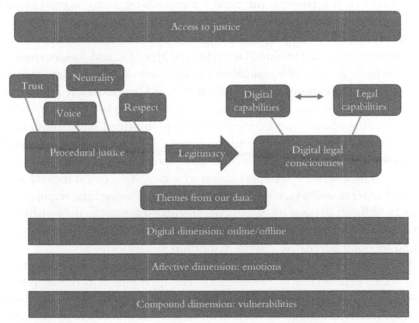

digital legal consciousness. Online and offline experiences may differ, and individuals may have varying levels of digital literacy and access to technology. This dimension may impact an individual's ability to access legal resources and participate in the justice system. In addition, the *compound dimension*, including vulnerabilities such as language barriers, disability or socioeconomic status, must also be considered. These factors can further limit an individual's ability to navigate legal processes, exacerbating the impact of the affective and digital dimensions. Next, we discuss these in turn.

Digital dimension: contrasting online and offline experiences

Our data showed that the digital dimension of a dispute resolution process can add extra layers of challenge and complexity. Without wanting to romanticize the face-to-face processes, we need to be mindful of what kind of justice system, or rather processes within a larger system, we are creating. A process people were used to doing offline is now online. The creators of the digital justice system must ask themselves if the service user has got the digital and legal capabilities to navigate the system. People need more assurance about the online process, which means that they need more procedural justice (to experience a process to be fair; see Chapter 2 and Chapter 6). However, there are questions about whether procedural justice can be effectively delivered online and how this would differ from face-to-face interactions. Service users must have the necessary capabilities to use online services, but service providers must also have the capabilities to deliver these services effectively. In this context, the need for procedural justice is even greater, as service users want assurance that they will be treated fairly. This will, in turn, provide access to justice and legitimacy for the online system.

Our data generated two main narratives of *Types 1, 2* and *3* users; these are *access to justice for vulnerable populations* and *legal processes affecting access to justice*. The first theme exposes unique challenges faced by vulnerable populations when accessing justice. This can include individuals with disabilities, those experiencing homelessness, or individuals with limited English proficiency. In Chapter 8 we discuss potential solutions to these challenges, such as providing targeted outreach efforts. The second theme shows how legal processes themselves can impact access to justice. This could include issues related to service delays, confusing legal jargon, or procedural requirements that are difficult to navigate. We discuss potential solutions to these challenges in Chapters 7 and 8, such as simplifying legal language signposting, or implementing alternative dispute resolution mechanisms.

Exclusion from the digital sphere and perceptions about online justice can mean that fairness is perceived differently online and offline (closely connected to questions of access and of capabilities). So, for a digital legal consciousness to emerge we need to understand how access to justice can be made fair online. We need to harness our capabilities to be able to access and navigate the online justice system. More work needs to be done to explore the relationship between these capabilities beyond how we set them out here in very general terms.

Type 1 service users have high levels of digital legal consciousness and are well-equipped to navigate the justice system. *Type 2* service users may have low digital legal consciousness but are motivated to learn and participate. *Type 3* service users lack the necessary digital legal consciousness but are open to learning and may be motivated by the promise of justice. *Type 4* service users have low levels of digital legal consciousness and may not be motivated to learn or participate in the justice system. This is a very problematic outlook and needs attention to avoid continuing to exclude people from access to online justice. In other words, we do not have a choice, but must develop our capabilities to be part of the online justice system. Or we may also come to recognize, in time, that online justice is not the way forward, at least in some areas. Alternatively, we might come to a more nuanced and informed idea of what does work well online, and what does not.

People's digital journeys require considerations of digital and legal needs. These can be different online than in a face-to-face interaction. At the most basic level we can see the challenges technology brings to the ability to access an online process. Our data showed that no matter how digitally capable a person is, when technology did not work, it added tension to an already fraught setting. In other words, digital systems can create their own difficulties and anxieties (distinct from, and in combination with, the legal aspect), so even people who prefer digital interactions for various reasons, or who are otherwise capable, are susceptible to these difficulties.

Affective dimension: emotions

Our data showed clearly that emotions play an important role in people's perceptions of accessing justice online. Again, in relation to *Types 1, 2* and *3*, our data showed examples of negative and positive perceptions of legal services which triggered a range of emotions. Negative perceptions of services can create a lack of trust in the legal system or a belief that legal services are not accessible or affordable. We discuss potential solutions to address these negative perceptions in Chapter 6, such as increasing transparency in the legal system or offering legal aid programmes. Positive perceptions of legal services can be promoted and can lead to increased access to justice and improved outcomes. We also explore potential strategies for promoting positive perceptions of legal services, such as outreach campaigns or community education initiatives. Having said this, people's negative perceptions of services are often well founded. Therefore, one aim has to be to encourage people to see that there may be some worthwhile possibility of justice so that they will consider taking action (for example, to challenge a sanction).

Our data clearly showed the disconnect between help-seekers' expectations about how the online process *ought* to go and what unfolds in reality. People's digital journeys can be influenced in a positive way through procedural justice. Procedural justice is crucial to establishing legitimacy in any justice system, and this is just as true in the digital space as it is offline. However, the nature of communication in digital spaces differs from that in offline contexts. It is not just *how* the communication is conducted, but also *what* is communicated that matters. In other words, the content of communication is just as important as its form, as it can make people feel like they are working with the process rather than having it work 'at' them. We found further that procedural justice has a significant impact on emotions, with procedural in/justice provoking negative emotions such as anger and frustration. Every pathway to justice brings with it an affective dimension. These emotions can be positive as well as negative. They depend on the context, the persons involved, on their complaint and on how it is dealt with.

Procedural justice helps to make the process feel more transparent. When people do not understand what happens next, or what is happening around them, this can create a sense of uncertainty about what is going on, which can then create anxiety, frustration and dissatisfaction. Our data suggest that treating people in interpersonally fair ways, giving people voice, and making fair decisions can make the process feel more transparent. In addition to this being a potential reason why procedural justice establishes legitimacy (people actually understand the process), procedural justice may also generate motive-based trust, which may then work as a heuristic through which people answer a difficult question (*do you understand the process?*) with an easy question (*do you trust the people driving the process and making decisions?*).

Emotions play an integral part in considerations of perceptions of fairness and the experience of an online dispute resolution process. Weaving elements of procedural justice into a process might support the help-seeker in experiencing a justice space that is accessible and supportive. However, emotions can make people vulnerable.

Compound dimension: vulnerabilities

Turning to *Type 4*, those who are not able to access the system, five narratives emerged predominately from vulnerable users that highlight that access to justice is not just about providing legal advice but also about providing support and guidance to manage legal needs. The themes are:

1. *Overcoming barriers to access to justice*, where we explore the various obstacles that individuals may face when trying to access justice, such as financial,

geographic and language barriers. We discuss potential solutions to overcome these barriers, such as providing improved access to support services or improving digital access to legal resources in Chapter 8.

2. *Providing support and guidance for legal needs* is about the importance of providing support and guidance to individuals who are navigating the legal system. This could include providing information about legal rights and resources, connecting individuals with legal representation, or offering emotional support throughout the process.

3. *Taking action for justice* covers the importance of taking action to address injustices in the system. This could include advocating for policy changes, participating in community organizing efforts, or pursuing legal action to hold individuals or organizations accountable for discriminatory practices.

4. *Navigating the impact of the pandemic on access to justice* is about the ways in which the COVID-19 pandemic has impacted access to justice. In Part III, we discuss how the pandemic has exacerbated existing inequalities and created new barriers to justice and explore potential solutions to address these challenges.

5. Finally, *promoting knowledge and awareness of tribunals and ombuds* is about the importance of promoting knowledge and awareness of tribunals and ombuds in public outreach initiatives.

This leads us to consider the concept of vulnerability. The vulnerabilities that service users bring with them can impact their experience of access to justice and of procedural justice. We can all be vulnerable when engaging with the justice system. While service providers may be doing the best they can, a lack of help-seekers' capabilities can undermine their ability to participate fully in the process. The degrees of vulnerability that people bring with them to the process varies, and our data showed that the online process can create or enhance vulnerabilities. It is impossible to grasp every individual's vulnerabilities and their variations. The four categories of digital legal consciousness are a starting point to capture varying levels of legal literacy, digital literacy, access to technology, and social and cultural capital.

Legal literacy is essential for individuals to understand their rights and to navigate legal systems and procedures (Grimes 2003; Goodwin and Maru 2017). Without legal literacy, individuals may not be aware of their rights or how to access justice. As mentioned before, digital literacy (Dobson and Willinsky 2009) and access to technology are also critical for access to justice, as many legal processes have moved online. Individuals need digital literacy skills and access to technology to access legal information and resources online, communicate with legal professionals, and participate in online proceedings. Social and cultural capital (Young and Billings 2020) can impact access to justice too. Those who have the means might have access to information and resources that can assist them to navigate the justice system more effectively. There are many complex factors that influence a person's ability to access advice, access an online justice system and to have an experience that they think of as fair.

Vulnerability can have a significant impact on a person's ability to access justice. Individuals who are vulnerable, such as those who are on a low income, have disabilities, or are part of marginalized communities, may face obstacles in accessing advice and legal services and in navigating the justice system. Addressing systemic biases and barriers, promoting awareness and education about resources, and assisting those who cannot access the online justice space in 'old-fashioned' and for them relatable ways, will commence the process of inclusion and access.

Further, hidden vulnerabilities (factors that are not immediately apparent) can pose barriers to accessing justice. These factors can include poverty, lack of education, mental health issues and social marginalization. Those who are affected by hidden vulnerabilities might have problems in navigating the justice system (for example, understanding the process, fear of retaliations, being prone to give up or not even try). These barriers can be addressed by simplifying the language around the processes, increasing awareness about rights, and thereby creating a more inclusive and equal system. Having said that, it is important to recognize that the various systems people have to navigate often actively exacerbate issues (for example, they may worsen someone's mental health or there may be punitive consequences for things that are beyond someone's control), so the systems themselves need to be less hostile and recognize that if someone is already vulnerable, they are going to struggle to engage in the ways those systems expect them to; to anticipate and accommodate that rather than holding it against them.

We found that access to online justice can have an impact on help-seekers' emotions in various ways. For example, online systems can help to alleviate negative emotions by providing a more accessible, efficient and cost-effective means of resolving disputes (Briggs 2016). Here, a sense of empowerment and control over one's situation, and the process to some extent, can lead to positive emotions. In contrast, the online justice system can also create negative emotions for help-seekers. As we outlined in Chapters 7 and 8 the online justice process can be overwhelming and frustrating, especially for those help-seekers who are not tech-savvy or those who have difficulty navigating the process and system. As we saw from our data, the online system does not provide the same level of human interaction and emotional support that can be found in the traditional face-to-face settings. The lack of this connection can lead to frustration and isolation. As the outcome of an online process is not always satisfactory, this can also lead to dissatisfaction. As shown in Chapter 6, it is important to provide a procedurally fair process so help-seekers experience the process as transparent and accountable, which enhances trust.

In British Columbia, user–centred design was applied to reimagine how to resolve disputes and to create an online dispute resolution system (Salter and Thompson 2017). This research also found, among other things, that the focus on better user experience in courts can lead to greater experiences of procedural justice (Hagan 2018: 220). One of the main points is that one needs to focus on the 'entire service journey of a person going through the legal system' (Hagan 2018: 237). This relates directly to our wider vision of access to justice; it is about assisting the help–seeker through a series of multiple tasks. There needs to be a 'sustained engagement' in the entire digital journey. The question remains though: to whom does this responsibility fall? Here, we return to thinking about how help–seekers' varying capabilities mean that this level of responsibility will vary. Those who are capable and knowledgeable about how to navigate the digital process will need less support throughout their digital journey than those who have fewer capabilities. Those who need a higher level of support would ideally be accompanied throughout their interaction with the system. This, however, brings us back to the reality of the scarce resources of advice providers and to the need to identify the necessary support for help–seekers early on.

As part of providing a justice system that is accessible, there needs to be consideration about how to evaluate new (or existing) processes to keep them aligned with their users' needs. Hagan suggests guiding principles that are aligned to procedural justice and can be applied to 'websites, apps, forms, flyers, posters, and other materials ... to see if these new things promote more consistency, full–journey support, system oversight, and mobile companionship' (2018: 222).

Hagan argues that a human-centred design that benefits access to justice through procedural justice can: (1) counter negative emotions; (2) reduce confusion and enhance perceived control; (3) strategically advise people; (4) allow for more efficient, simple experiences; and (5) give respect and dignity through the process. All these elements resonate with our study, and we address them through the lens of digital legal consciousness in Figure 9.1.

We have presented a comprehensive model that elucidates how procedural justice fosters legitimacy within the online justice system. This model encourages us to contemplate access to justice in a more comprehensive and inclusive manner, transcending the traditional legal boundaries. By emphasizing the necessary skills required by help-seekers and addressing the potential challenges they may encounter, we aim to facilitate optimal access to online justice. Our work provides an initial framework for reimagining access to justice, promoting a more nuanced and accommodating approach.

We have also found that procedural justice matters in the online justice context, contributing to exciting emerging spaces for research (Mentovich et al 2023). Further, in the context of the pandemic and the digitalization of the justice system, our data provided us with examples of challenges alongside the benefits and promises of online advice and dispute resolution systems. Our data suggested, however, it remains essential to keep open those pathways that allow people to access the advice and justice landscape face-to-face or via the telephone. At this point in time there remains a significant number of the population that are not able to access a purely online system. What this means in our context of the AJS is that the speed at which the pandemic forced an ongoing, slow and careful digitalization process to be stress-tested revealed the obstacles that do exist. Now, there is an opportunity to overcome them. One route for this might be to think about a better joined-up AJS, and one step in that direction is to return to the discission about ombuds and tribunal partnerships, the place from which this project was inspired.

In sum, to ensure access to justice in the online justice system we need to think about *access* in a broad and inclusive sense. To help with this, it is important to understand how people engage with, and relate to, the digital (justice) system. An individual's digital legal consciousness provides answers to this. It is essential to consider the dimensions discussed earlier, including the affective, digital and compound dimensions. This wider understanding can help identify barriers to access and inform strategies to improve access to justice. A more nuanced understanding of digital legal consciousness requires ongoing research and collaboration between legal practitioners, policymakers and technology experts. By recognizing the complexity of individual experiences, we can work towards creating a more equitable and accessible justice system for all.

References

Briggs, Lord Justice (2016) 'Civil courts structure review: final report', Judiciary of England and Wales. www.judiciary.uk/wp-content/uploads/2016/07/civil-courts-structure-review-final-report-jul-16-final-1.pdf

Creutzfeldt, N. (2021) 'Towards a digital legal consciousness?', *European Journal of Law and Technology* 12(3). https://ejlt.org/index.php/ejlt/article/view/816

Creutzfeldt, N. and Bradford, B. (2016) 'Dispute resolution outside of courts: procedural justice and decision acceptance among users of ombuds services in the UK', *Law & Society Review* 50(4): 985–1016.

Denvir, C. and Selvarajah, A.D. (2022) 'Safeguarding access to justice in the age of the online court', *Modern Law Review* 85(1): 25–68. doi:10.1111/1468-2230.12670

Denvir, C., Sutherland, C., Selvarajah, A.D., Balmer, N. and Pleasence, P. (2021) 'Access to online courts: exploring the relationship between legal and digital capability'. https://papers.ssrn.com/sol3/papers.cfm?abstract_id=3838153

Dobson, T. and Willinsky, J. (2009) 'Digital literacy', in *The Cambridge Handbook of Literacy*, Cambridge: Cambridge University Press, 286–312.

Goodwin, L. and Maru, V. (2017) 'What do we know about legal empowerment? Mapping the evidence', *Hague Journal on the Rule of Law* 9: 157–194.

Grimes, R. (2003) 'Legal literacy, community empowerment and law schools: some lessons from a working model in the UK', *The Law Teacher* 37(3): 273–284.

Hagan, M.D. (2018) 'A human-centered design approach to access to justice: generating new prototypes and hypotheses for intervention to make courts user-friendly', *Indiana Journal of Law and Social Equality* 6(2): article 2. www.repository.law.indiana.edu/ijlse/vol6/iss2/2

Hou, Y., Lampe, C., Bulinski, M. and Prescott, J.J. (2017) 'Factors in fairness and emotion in online case resolution systems', *Proceedings of the 2017 CHI Conference on Human Factors in Computing Systems* (May): 2511–2522.

Mentovich, A., Prescott, J.J. and Rabinovich-Einy, J.J. (2023) 'Legitimacy and online proceedings: procedural justice, access to justice, and the role of income', *Law & Society Review* 57: 189–213.

Mulcahy, L. and Rowden, E. (2020) *The Democratic Courthouse: A Modern History of Design, Due Process and Dignity*, London: Routledge.

Salter, S. and Darin Thompson, D. (2017) 'Public-centred civil justice redesign: a case study of the British Columbia Civil Resolution Tribunal', *McGill Journal of Dispute Resolution* 3(2016–2017): 113–136.

Schmitz, A.J. (2019) 'Measuring "access to justice" in the rush to digitize', *Fordham Law Review* 88: 2381.

Young, K.M. and Billings, K.R. (2020) 'Legal consciousness and cultural capital', *Law & Society Review* 54(1): 33–65.

Examples of Vignettes

Contents

1. Vignette example: a fair online tribunal process

Marta is 40 years old, a single mother of two living in social housing, who has been struggling to pay her housing costs and rent since COVID-19 hit and caused her to lose her job. She was able to continue with most of her regular payments. She has now found a part-time job and been able to clear some of her arrears and pay the ongoing rent. However, her landlord has now given notice that the rent is to be increased, and Marta cannot afford it. The advice organization Marta contacts helps her figure out what to do next. She can take her problem to the Property Chamber which is a tribunal. The Property Chamber handles applications, appeals and references relating to disputes over property and land. Residential property disputes that they handle include rent increases for 'fair' or 'market' rates.

Marta appeals to the Property Chamber for a decision about the proposed rent increase. Marta needs to fill in a form to make the appeal. She downloads the form from the HM Courts & Tribunals Service website. She fills it in and posts it. Usually there is a fee of £100 to pay, but there is a 'fee waiver' available for those who need it. The advice organization helps Marta get the fee waiver, so she does not have to pay the £100.

Marta waits to hear back from the Property Chamber. The Property Chamber checks Marta's form and the extra attachments she sent with it. It then gets back to Marta with a timetable for her case, the date of her hearing, and some extra information about the hearing. *It also provides Marta with a leaflet about her legal rights and their procedures, and advises that she can read more on their website [PJ (procedural justice): understanding legal rights & procedures]. The hearing is arranged to take place by video [online hearing]. Marta is confident with computers, so she is happy with this. She is told that she can go to the tribunal in person if she would prefer.*

Marta attends the hearing at the tribunal online. The people who are in charge of the hearing and who will decide the case are there. They are called the 'panel' and are made up of one judge and two other people who know about housing issues. A local authority representative is also at the hearing. *The judge commences the hearing, addressing Marta directly: "Good morning. Thank you for being here today and presenting your side of the story. I apologize for the wait time this morning. Each case is important to me, and we will work together to get through today's calendar as quickly as possible, while giving each case the time it needs. Let's get started" [PJ: respect]. The judge recites the basic rules and format of the court proceedings and written procedures are also posted in the chat function to reinforce understanding. As part of this, the judge tells everybody present to put their phones on silent to ensure the hearing runs smoothly [PJ: feeling like the process is transparent and applied the same way for all]. Then the judge tells Marta, "Having looked through the documentation for this case, I understand your frustration. Please explain to me what happened and I will try to help" [PJ: voice]. Marta expresses her viewpoint that she feels the rent asked of her is unfair and that she cannot afford it. The judge engages with Marta and tells her that he will take on board everything said today before he makes his decision. Marta is assured by the judge that she will receive the decision in writing in due course.*

The Property Chamber writes to Marta to tell her its decision and sends a copy of the decision to the landlord. The Property Chamber decides that the increased rent is more than it would be for similar properties and that the increase would be unreasonable. It decides her current rent is accurate and Marta, therefore, does not need to pay the increased rent.

2. Vignette example: an unfair online tribunal process

Marta is 40 years old, a single mother of two living in social housing, who has been struggling to pay her housing costs and rent since COVID-19 hit and caused her to lose her job. She was able to continue with most of her regular payments. She has now found a part-time job and been able to clear some of her arrears and pay the ongoing rent. However, her landlord has now given notice that the rent is to be increased, and Marta cannot afford it. The advice organization Marta contacts helps her figure out what to do next. She can take her problem to the Property Chamber which is a tribunal. The Property Chamber handles applications, appeals and references relating to disputes over property and land. Residential property disputes that they handle include rent increases for 'fair' or 'market' rates.

Marta appeals to the Property Chamber for a decision about the proposed rent increase. Marta needs to fill in a form to make the appeal. She downloads the form from the HM Courts & Tribunals Service website. She fills it in and posts it. Usually there is a fee of £100 to pay, but there is a 'fee waiver' available for those who need it. The advice organization helps Marta get the fee waiver, so she does not have to pay the £100.

Marta waits to hear back from the Property Chamber. The Property Chamber checks Marta's form and the extra attachments she sent with it. It then gets back to Marta with a timetable for her case, the date of her hearing, and some extra information about the hearing. *It does not provide Marta with any information about her legal rights and procedures; nor does it tell her where she can read more about the Property Chamber [PJ: (not) understanding legal rights & procedures]. The hearing is arranged to take place by video [online hearing]. Marta is confident with computers, so she is happy with this. She is told that she can go to the tribunal in person if she would prefer.*

Marta attends the hearing at the tribunal. The people who are in charge of the hearing and who will decide the case are there. They are called the 'panel', and are made up of one judge and two other people who know about housing issues. A local authority representative is also at the hearing. *The judge commences the hearing without addressing Marta directly: "Good morning. We are late. Let's get started." [PJ: (dis)respect]. The judge rushes through the basic rules and format of the court proceedings but written procedures are not posted in the chat function to reinforce understanding. The judge tells Marta, but no one else on the call, to put her phone on silent to ensure the hearing runs smoothly [PJ: (not) feeling like the process is transparent and applied the same way for all]. The judge starts by saying, "Having read through all the documentation, I don't think we need to take any more evidence from the parties" [PJ: (lack of) voice]. Marta feels it is unfair that she is not invited to express her viewpoint. The judge does not engage with Marta, nor does he tell her that he will take on board everything said today before he makes his decision. Marta leaves the room without assurance by the judge that she will receive the decision in writing in due course.*

The Property Chamber writes to Marta to tell her its decision and sends a copy of the decision to the landlord. The Property Chamber decides that the increased rent is more than it would be for similar properties and that the increase would be unreasonable. It decides her current rent is accurate and Marta, therefore, does not need to pay the increased rent.

3. Vignette example: a fair offline tribunal process

Marta is 40 years old, a single mother of two living in social housing, who has been struggling to pay her housing costs and rent since COVID-19 hit and caused her to lose her job. She was able to continue with most of her regular payments. She has now found a part-time job and been able to clear some of her arrears and pay the ongoing rent. However, her landlord has now given notice that the rent is to be increased, and Marta cannot afford it. The advice organization Marta contacts helps her figure out what to do next. She can take her problem to the Property Chamber which is a tribunal. The Property Chamber handles applications, appeals and references relating to disputes over property and land. Residential property disputes that they handle include rent increases for 'fair' or 'market' rates.

Marta appeals to the Property Chamber for a decision about the proposed rent increase. Marta needs to fill in a form to make the appeal. She downloads the form from the HM Courts & Tribunals Service website. She fills it in and posts it. Usually there is a fee of £100 to pay, but there is a 'fee waiver' available for those who need it. The advice organization helps Marta get the fee waiver, so she does not have to pay the £100.

Marta waits to hear back from the Property Chamber. The Property Chamber checks Marta's form and the extra attachments she sent with it. It then gets back to Marta with a timetable for her case, the date of her hearing, and some extra information about the hearing. *It provides Marta with a leaflet about her legal rights and procedures and tells her that she can read more on their website [PJ: understanding legal rights & procedures]. Marta is told to attend the tribunal in person [offline hearing]. Marta is happy with this. But she is told that the hearing can be arranged to take place by video if she prefers.*

Marta attends the hearing at the tribunal. The people who are in charge of the hearing and who will decide the case are there. They are called the 'panel', and are made up of one judge and two other people who know about housing issues. A local authority representative is also at the hearing. *The judge commences the hearing, making eye contact with Marta: "Good morning. Thank you for being here today and presenting your side of the story. I apologize for the wait time this morning. Each case is important to me, and we will work together to get through today's calendar as quickly as possible, while giving each case the time it needs. Let's get started" [PJ: respect]. The judge recites the basic rules and format of the court proceedings and written procedures are also posted in the courtroom to reinforce understanding. As part of this, the judge tells everybody present to put their phones on silent to ensure the hearing runs smoothly [PJ: feeling like the process is transparent and applied the same way for all]. Then the judge tells Marta, "I understand your frustration. Please explain to me what happened and I will try to help" [PJ: voice]. Marta expresses her viewpoint that she feels the rent asked of her is unfair and that she cannot afford it. The judge engages with Marta and tells her that he will take on board everything said today before he makes his decision. Marta is assured by the judge that she will receive the decision in writing in due course.*

The Property Chamber writes to Marta to tell her its decision and sends a copy of the decision to the landlord. The Property Chamber decides that the increased rent is more than it would be for similar properties and that the increase would be unreasonable. It decides her current rent is accurate and Marta, therefore, does not need to pay the increased rent.

4. Vignette example: an unfair offline tribunal process

Marta is 40 years old, a single mother of two living in social housing, who has been struggling to pay her housing costs and rent since COVID-19 hit and caused her to lose her job. She was able to continue with most of her regular payments. She has now found a part-time job and been able to clear some of her arrears and pay the ongoing rent. However, her landlord has now given notice that the rent is to be increased, and Marta cannot afford it. The advice organization Marta contacts helps her figure out what to do next. She can take her problem to the Property Chamber which is a tribunal. The Property Chamber handles applications, appeals and references relating to disputes over property and land. Residential property disputes that they handle include rent increases for 'fair' or 'market' rates.

Marta appeals to the Property Chamber for a decision about the proposed rent increase. Marta needs to fill in a form to make the appeal. She downloads the form from the HM Courts & Tribunals Service website. She fills it in and posts it. Usually there is a fee of £100 to pay, but there is a 'fee waiver' available for those who need it. The advice organization helps Marta get the fee waiver, so she does not have to pay the £100.

Marta waits to hear back from the Property Chamber. The Property Chamber checks Marta's form and the extra attachments she sent with it. It then gets back to Marta with a timetable for her case, the date of her hearing, and some extra information about the hearing. *It does not provide Marta with any information about her legal rights and procedures; nor does it tell her where she can read more about the Property Chamber [PJ: (not) understanding legal rights & procedures]. Marta is told to attend the tribunal in person [offline hearing]. Marta is happy with this, but she is told that the hearing can be arranged to take place by video if she prefers.*

Marta attends the hearing at the tribunal. The people who are in charge of the hearing and who will decide the case are there. They are called the 'panel', and are made up of one judge and two other people who know about housing issues. A local authority representative is also at the hearing. *The judge commences the hearing without making eye contact with Marta: "Good morning. We are late. Let's get started" [PJ: (dis)respect]. The judge rushes through the basic rules and format of the court proceedings but written procedures are not posted in the courtroom to reinforce understanding. The judge tells only Marta to put her phone on silent to ensure the hearing runs smoothly [PJ: (not) feeling like the process is transparent and applied the same way for all]. The judge starts by saying, "Having read through all the documentation, I don't think we need to take any more evidence from the parties" [PJ: (lack of) voice]. Marta feels it is unfair that she is not invited to express her viewpoint. The judge does not engage with Marta, nor does he tell her that he will take on board everything said today before he makes his decision. Marta leaves the room without assurance by the judge that she will receive the decision in writing in due course.*

The Property Chamber writes to Marta to tell her its decision and sends a copy of the decision to the landlord. The Property Chamber decides that the increased rent is more than it would be for similar properties and that the increase would be unreasonable. It decides her current rent is accurate and Marta, therefore, does not need to pay the increased rent.

5. Vignette example: a fair online ombuds process

Marta is 40 years old, a single mother of two living in social housing, who has been struggling to pay her housing costs and rent since COVID-19 hit and caused her to lose her job. She was able to continue with most of her regular payments. She has now found a part-time job and been able to clear some of her arrears and pay the ongoing rent. However, her landlord has now given notice that the rent is to be increased, and Marta cannot afford it. The advice organization Marta contacts helps her figure out what to do next. She can take her problem to the Housing Ombudsman. The Housing Ombudsman Service is set up by law to look at complaints about housing organizations. Residents and landlords can contact the Ombudsman at any time for support in helping to resolve a dispute.

She submits her form online to the Housing Ombudsman. Marta can also go through her MP, a local councillor or a tenant panel to reach the Ombudsman but she decides to skip this step and go straight to the Ombudsman.

Marta waits for a decision from the Ombudsman. *The complaint handlers provide Marta with a leaflet about her legal rights and their procedures, and tell her where she can read more on their website [PJ: understanding legal rights & procedures]. Contact with the complaint handler is arranged to take place by email [online]. But Marta is told that she can talk to the complaint handler on the phone if she would prefer.*

After the complaint handlers at the Ombudsman have checked Marta's complaint, prior to passing on the complaint to the Housing Ombudsman for a full investigation, the investigation team works with Marta via email to resolve the dispute by asking Marta for more detailed information about what happened. *The investigation team writes to Marta, directly addressing her: "Good morning. Thank you for presenting your side of the story. We need some more detailed information about what happened so please take the time to respond to our enquiries" [PJ: respect]. They go on to say, "Having looked through the documentation for this case, we understand your frustration. Please explain to us what happened in more detail and we will try to help" [PJ: voice]. "It is important that you respond fully to any questions put to you and provide evidence to support what you say. However, please only send the evidence we have asked for, unless you feel there is a crucial piece of information we should also see. If you are not sure whether to include a piece of information, please contact the investigator directly via email to discuss. For fairness, we aim to share any evidence that we rely on in reaching a decision. We will not share any information about a third party or that is confidential for another reason" [PJ: feeling like the process is transparent and applied the same way for all]. Marta writes back to the investigation team expressing her viewpoint that she feels the rent asked of her is unfair and that she cannot afford it. The investigator engages with Marta via email and tells her that they will take on board everything said today before the Ombudsman makes their decision. Marta is assured by the investigation team that she will receive the decision in writing in due course.*

After a few months, the Housing Ombudsman makes a decision. It makes suggestions to the landlord to resolve the issue. It encourages the landlord to give Marta longer to pay the arrears and to reverse the increase in rent to the original rent. In this example, Marta has a successful outcome, and she is happy that she no longer needs to pay a higher rent.

6. Vignette example: an unfair online ombuds process

Marta is 40 years old, a single mother of two living in social housing, who has been struggling to pay her housing costs and rent since Covid-19 hit and caused her to lose her job. She was able to continue with most of her regular payments. She has now found a part-time job and been able to clear some of her arrears and pay the ongoing rent. However, her landlord has now given notice that the rent is to be increased, and Marta cannot afford it. The advice organization Marta contacts helps her figure out what to do next. She can take her problem to the Housing Ombudsman. The Housing Ombudsman Service is set up by law to look at complaints about housing organizations. Residents and landlords can contact the Ombudsman at any time for support in helping to resolve a dispute.

She submits her form online to the Housing Ombudsman. Marta can also go through her MP, a local councillor or a tenant panel to reach the Ombudsman but she decides to skip this step and go straight to the Ombudsman.

Marta waits for a decision from the Ombudsman. *The complaint handlers do not provide Marta with any information about her legal rights and procedures; nor do they tell her where she can read more about the Housing Ombudsman [PJ: (not) understanding legal rights & procedures]. Contact with the complaint handler is arranged to take place by email [online]. But Marta is told that she can talk to the complaint handler on the phone if she would prefer.*

After the complaint handlers at the Ombudsman have checked Marta's complaint, prior to passing on the complaint to the Housing Ombudsman for a full investigation, the investigation team works with Marta via email to resolve the dispute by asking her for more detailed information about what happened. *The investigation team writes to Marta, without directly addressing her: "Good morning. We don't have much time to work on your complaint" [PJ: (dis) respect]. They go on to say, "Having read through all the documentation, I don't think we need to take any more evidence from the parties" [PJ: (lack of) voice]. "Normally we ask complainants for detailed information about what happened and ask them to take the time to respond to our enquiries, but I don't think that's necessary for this case" [PJ: (not) feeling like the process is transparent and applied the same way for all].*

Marta writes back to say that she feels it is unfair that she is not invited to express her viewpoint. The investigator does not engage with Marta via email, nor does he tell her that they will take on board everything said today before the Ombudsman makes their decision. Marta is assured by the investigation team that she will receive the decision in writing in due course.

After a few months, the Housing Ombudsman makes a decision. It makes suggestions to the landlord to resolve the issue. It encourages the landlord to give Marta longer to pay the arrears and to reverse the increase in rent to the original rent. In this example, Marta has a successful outcome and she is happy that she no longer needs to pay a higher rent.

7. Vignette example: a fair offline ombuds process

Marta is 40 years old, a single mother of two living in social housing, who has been struggling to pay her housing costs and rent since COVID-19 hit and caused her to lose her job. She was able to continue with most of her regular payments. She has now found a part-time job and been able to clear some of her arrears and pay the ongoing rent. However, her landlord has now given notice that the rent is to be increased, and Marta cannot afford it. The advice organization Marta contacts helps her figure out what to do next. She can take her problem to the Housing Ombudsman. The Housing Ombudsman Service is set up by law to look at complaints about housing organizations. Residents and landlords can contact the Ombudsman at any time for support in helping to resolve a dispute.

She submits her form online to the Housing Ombudsman. Marta can also go through her MP, a local councillor or a tenant panel to reach the Ombudsman but she decides to skip this step and go straight to the Ombudsman.

Marta waits for a decision from the Ombudsman. *The complaint handlers provide Marta with a leaflet about her legal rights and their procedures and tells her where she can read more on their website [PJ: understanding legal rights & procedures].*

Contact with the complaint handler is arranged to take place by phone [offline]. But Marta is told that she can talk to the complaint handler via email if she would prefer.

After the complaint handlers at the Ombudsman have checked Marta's complaint, prior to passing on the complaint to the Housing Ombudsman for a full investigation, the investigation team works with Marta on the phone to resolve the dispute by asking Marta for more detailed information about what happened. *The investigation team says to Marta, directly addressing her: "Good morning. Thank you for presenting your side of the story. We need some more detailed information about what happened so please take the time to respond to our enquiries" [PJ: respect]. They go on to say, "Having looked through the documentation for this case, we understand your frustration. Please explain to us what happened in more detail and we will try to help" [PJ: voice]. "It is important that you respond fully to any questions put to you and provide evidence to support what you say. However, please only send the evidence we have asked for, unless you feel there is a crucial piece of information we should also see. If you are not sure whether to include a piece of information, please contact the investigator directly to discuss on the phone. For fairness, we aim to share any evidence that we rely on in reaching a decision. We will not share any information about a third party or that is confidential for another reason" [PJ: feeling like the process is transparent and applied the same way for all]. Marta expresses her viewpoint that she feels the rent asked of her is unfair and that she cannot afford it. The investigator engages with Marta and tells her that they will take on board everything said today before the Ombudsman makes their decision. Marta is assured by the investigation team that she will receive the decision in writing in due course.*

After a few months, the Housing Ombudsman makes a decision. It makes suggestions to the landlord to resolve the issue. It encourages the landlord to give Marta longer to pay the arrears and to reverse the increase in rent to the original rent. In this example, Marta has a successful outcome, and she is happy that she no longer needs to pay a higher rent.

8. Vignette example: an unfair offline ombuds process

Marta is 40 years old, a single mother of two living in social housing, who has been struggling to pay her housing costs and rent since COVID-19 hit and caused her to lose her job. She was able to continue with most of her regular payments. She has now found a part-time job and been able to clear some of her arrears and pay the ongoing rent. However, her landlord has now given notice that the rent is to be increased, and Marta cannot afford it. The advice organization Marta contacts helps her figure out what to do next. She can take her problem to the Housing Ombudsman. The Housing Ombudsman Service is set up by law to look at complaints about housing organizations. Residents and landlords can contact the Ombudsman at any time for support in helping to resolve a dispute.

She submits her form online to the Housing Ombudsman. Marta can also go through her MP, a local councillor or a tenant panel to reach the Ombudsman but she decides to skip this step and go straight to the Ombudsman.

Marta waits for a decision from the Ombudsman. *The complaint handlers do not provide Marta with any information about her legal rights and procedures; nor do they tell her where she can read more about the Housing Ombudsman [PJ: (not) understanding legal rights & procedures]. Contact with the complaint handler is arranged to take place by phone [offline]. But Marta is told that she can talk to the complaint handler via email if she would prefer.*

After the complaint handlers at the Ombudsman have checked Marta's complaint, prior to passing on the complaint to the Housing Ombudsman for a full investigation, the investigation team works with Marta on the phone to resolve the dispute by asking Marta for more detailed information about what happened. *The investigator says to Marta, without directly addressing her: "Good morning. We are late. Let's get started" [PJ: (dis)respect]. They go on to say, "Having read through all the documentation, I don't think we need to take any more evidence from the parties" [PJ: (lack of) voice]. "Normally we ask complainants for detailed information about what happened and ask them to take the time to respond to our enquiries, but I don't think that's necessary for this case" [PJ: (not) feeling like the process is transparent and applied the same way for all]. Marta feels it is unfair that she is not invited to express her viewpoint. The investigator does not engage with Marta nor does he tell her that they will take on board everything said today before the Ombudsman makes their decision. Marta puts down the phone without assurance by the investigation team that she will receive the decision in writing in due course.*

After a few months, the Housing Ombudsman makes a decision. It makes suggestions to the landlord to resolve the issue. It encourages the landlord to give Marta longer to pay the arrears and to reverse the increase in rent to the original rent. In this example, Marta has a successful outcome and she is happy that she no longer needs to pay a higher rent.

Note
When we conducted the experiment these vignettes had illustrations, they have been removed here. The text in italics shows our procedural justice manipulations for the eight vignettes.

Index

References to figures are in *italics*; tables in **bold**

Printed and bound by CPI Group (UK) Ltd, Croydon, CR0 4YY

27/10/2024

14580557-0005